Quantum Physics and the Power of the Mind

A Clear and Practical Guide to Mechanics, Quantum Thinking, and the Law of Attraction to Shape Reality and Manifest What Truly Matters

Samantha Ballantine

Table of Contents

Introduction

There's something about the human mind that science still hasn't pinned down. We can explain neurons, chemicals, and patterns of activity in the brain, but awareness itself — the fact that we know we exist — remains outside of full explanation.

At the same time, quantum physics has shown us that the fabric of reality behaves in strange, unpredictable ways at the smallest scales. Particles don't follow simple rules. They exist in multiple states, respond to observation, and don't settle into clear outcomes until something interacts with them.

Some theories suggest these two unknowns — consciousness and quantum behavior — might be connected. Not as metaphor, but as fact. The idea isn't universally accepted, but it's also not easy to dismiss. The observer effect, entanglement, and non-locality challenge the idea of a world made only of matter. They leave room for the mind to have a role.

This book looks at the possibility that thought, awareness, and focused mental energy can influence physical outcomes. It doesn't promise easy answers or instant results. It doesn't reject science, nor does it blindly follow it. The goal is to follow what we do know, question what we assume, and look closer at what experience tells us.

Some of what follows is supported by evidence. Some of it is speculative. But all of it is connected to a larger question: are we just passive observers in a mechanical universe, or are we part of something more dynamic — where mind and matter interact?

This isn't about belief. It's about paying attention.

Chapter 1: Quantum Consciousness

There are moments when thought stops. No inner voice, no mental images — just a clear sense of being. That state is not easy to explain, but most people have felt it at least once. It's what remains when the noise fades. Some call it awareness, others call it presence. In this book, we'll call it consciousness.

This isn't about beliefs, and it's not about faith. It's about something simple and immediate: the fact that we experience. Before language, before memory, before identity — there's a sense of being aware. That's the starting point.

Quantum physics, on the other hand, studies particles so small they can't be seen, only measured. Their behavior is strange: unstable, unpredictable, and sensitive to interaction. One of the most surprising things scientists discovered is that observation affects outcome. That single fact opens a door — a quiet but serious suggestion that the act of being aware might not be separate from the physical world.

This chapter looks at that possibility. Not with the goal of proving anything, but with the goal of asking better questions. What is consciousness? Where does it come from? And what does it have to do with a universe made of particles that respond to being observed?

What Is Consciousness?

Consciousness is the one thing you can't observe from the outside, yet it's the one thing you're most certain exists. You can doubt every belief, reject every thought, question every feeling — but you can't deny the fact that you're aware. That fact is what this chapter begins with.

Most modern theories treat consciousness as a byproduct of brain activity. In that view, once neurons reach a certain level of complexity, awareness appears — like steam rising from boiling water. But there's a problem: no one has explained how this happens. We can measure the brain's activity. We can see what areas light up during speech, fear, memory, or decision-making. But those scans don't explain *why* there's a subjective experience tied to that activity.

You can program a machine to recognize a face. You can train it to respond to speech. But it doesn't feel anything. It has no inner life. No hunger. No sense of time. No sense of "I am." Consciousness includes that internal presence — the first-person experience of being.

Philosophers call this "the hard problem" of consciousness: explaining *how* physical processes in the brain produce subjective experience. A machine can simulate thinking. It can process data and respond. But as far as we know, it doesn't experience anything. It doesn't *know* it's doing anything. That difference — between function and awareness — is where the problem lies.

Some theories suggest we're looking in the wrong place. Instead of trying to explain how the brain creates consciousness, maybe we should ask a different question: *What if consciousness isn't created at all? What if it's a basic feature of the universe, like space or time?* In that case, the brain wouldn't produce awareness — it would receive or filter it.

This isn't a new idea. Some ancient systems of thought describe consciousness as the foundation of reality, not just a product of biology. In those views, the mind doesn't come *from* matter — matter arises *within* mind. More recently, some scientists and philosophers have proposed similar models. The term for one such view is panpsychism — the idea that consciousness is a basic feature of all things, even if in primitive form.

There's another perspective called the consciousness field hypothesis. According to this model, consciousness exists as a field — like gravity or magnetism — and living beings are tuned into it. The brain doesn't create consciousness; it acts more like a receiver or a lens. This would explain why no single part of the brain has been identified as the source of awareness. It would also explain why in some rare cases, individuals with damaged or incomplete brains still show signs of normal — even high — intelligence.

If this sounds like science fiction, it's worth pointing out that several leading physicists and neuroscientists take these ideas seriously. The reason isn't that the theories are proven — they aren't — but that the standard explanations fail to account for the experience of being. Measuring brainwaves doesn't explain awareness. Counting neurons doesn't account for the sense of self.

There's also a practical reason why this matters. If consciousness is not just a side-effect of the brain, but a force or field in its own right, then thoughts and awareness could interact with the world in ways science doesn't yet understand. That doesn't mean thoughts control reality. But it does raise the possibility that observation — and attention — might have a real effect on physical outcomes.

This connects to findings in quantum physics, especially the observer effect. When a particle behaves differently depending on whether or not it's observed, it challenges the idea that observation is passive. If consciousness changes outcomes in quantum systems, even in subtle ways, then it might be playing a role in shaping events at the most basic level of matter.

What we call "consciousness" may not be limited to the brain, or even to living organisms. It might be a property that runs through the structure of everything — just usually unnoticed. The reason we experience it so clearly may be that the human brain, through its complexity, is especially good at focusing and amplifying it.

So what is consciousness? It's the question behind all other questions. Before science, before thought, before language — there's the sense of being aware. And until we understand that, everything else we think we know about the mind, the brain, or reality itself may be incomplete.

Basic Principles of Quantum Physics

Quantum physics studies how the universe behaves at the smallest possible scales — smaller than atoms, smaller than protons or electrons. At this level, the rules that govern the physical world start to break down, or at least shift into something unfamiliar. What looks solid and predictable at our level becomes unstable, uncertain, and deeply strange.

Classical physics, the kind used to build bridges or launch rockets, is based on logic and cause-effect relationships. If you know the conditions of a system — the speed, mass, and force — you can predict how it will behave. Newton's laws describe how planets orbit, how objects fall, and how collisions work. But those laws don't apply in the same way when you shrink things down to the level of subatomic particles.

At the quantum scale, particles don't always act like individual pieces of matter. Sometimes they act like waves — spread out, continuous, and able to interfere with each other. This is called wave-particle duality. An electron, for example, is not just a point flying through space. It behaves like a wave of possibilities. It has no fixed position until something forces it to take one.

This leads to a central concept in quantum mechanics: superposition. A particle doesn't have to be in one place or state. It can exist in multiple states at once. For example, a particle might be spinning clockwise and counterclockwise at the same time, or be in two positions at once. These possibilities only "collapse" into a single outcome when measured or observed.

The idea of superposition was illustrated famously by the thought experiment known as Schrödinger's cat. Imagine a cat in a box, along with a mechanism that has a 50/50 chance of releasing poison depending on the state of a quantum particle. Until someone opens the box and observes, the particle is in a superposition — both decayed and not decayed. So the cat is both alive and dead. Only observation brings it into a definite state. This was meant to show the absurdity of applying quantum rules to large objects, but it also highlights how observation seems to play a role in outcomes.

Another core principle is quantum entanglement. When two particles interact and become entangled, they remain connected no matter how far apart they are. If you measure one particle's state, the other particle's state responds instantly — even if it's on the other side of the galaxy. Einstein called this "spooky action at a distance" because it seemed to violate the rule that nothing can travel faster than light. But experiments have repeatedly confirmed that entanglement is real.

Entanglement suggests that space and distance may not work the way we think they do — at least not in the quantum world. It implies that there may be deep, invisible connections between objects, and that information might be shared in ways that bypass normal channels. This has led to serious questions about whether the universe is truly local, or if everything is part of a connected whole.

Another important principle is quantum uncertainty, first described by Werner Heisenberg. The idea is that you can't know both the position and momentum of a particle at the same time with complete accuracy. The more precisely you measure one, the less precisely you can measure the other. This isn't just a technical limitation — it's a fundamental part of nature at the quantum level.

The act of measurement itself introduces change. When you try to find out where a particle is, your tools disturb it. But even beyond that, there's something about the structure of reality that prevents full certainty. This is where the concept of the observer effect enters the picture.

In many quantum experiments, especially the double-slit experiment, the mere act of observing changes the behavior of particles. Without observation, particles behave like waves — passing through both slits at once, forming interference patterns. With observation, they act like individual particles, going through one slit or the other. This shift happens simply because the system is being measured.

These are not metaphors or mystical ideas. They're results from well-documented, repeatable experiments. And yet they point to something that challenges the normal view of reality. In the classical view, the world exists in a fixed state, whether we look or not. In the quantum view, observation helps shape what happens. This doesn't mean reality doesn't exist without us — but it does mean that measurement and awareness play a role in how outcomes take form.

This is why some physicists and philosophers have suggested that consciousness may be involved in quantum behavior. It's not about bending reality with thought. It's about recognizing that the structure of matter itself seems to respond to observation — and that the observer is not separate from the system being observed.

These principles — wave-particle duality, superposition, entanglement, uncertainty, and the observer effect — are not just technical curiosities. They challenge the basic way we think about matter, space, and time. And they open a door to the possibility that the mind, through its role in observation and attention, may be part of the same picture.

The Role of the Observer

In everyday life, observing something doesn't seem to change it. You can look at a tree, a cup, or a cloud without affecting its behavior. But in quantum physics, things work differently. At the smallest scales of matter, the act of observing – or more precisely, measuring – changes what's being observed. This is not just a side effect of using tools or disturbing a system. It appears to be a built-in feature of how reality operates at the quantum level.

The experiment that made this clear is called the double-slit experiment, and it has been repeated in many forms since the early 20th century. Here's how it works. A single particle, like an electron or photon, is fired at a barrier with two narrow slits. On the other side of the barrier is a screen that records where the particle lands. If no one measures which slit the particle goes through, it behaves like a wave. It passes through *both* slits at once, creating an interference pattern on the screen – the kind you'd expect from overlapping waves.

But if a measuring device is placed at the slits to detect which path the particle takes, the outcome changes. The interference pattern disappears, and the particle behaves like a single object, going through one slit or the other. The wave behavior is gone. The only thing that changed was the presence of a measurement – observation. Somehow, observing the particle forces it to "choose" a state.

This is called the observer effect. It's one of the most unusual and debated aspects of quantum theory. It suggests that particles don't have fixed properties until they're measured. Before observation, they exist in a state of probability – a wave of potential outcomes. Once observed, they settle into a single reality. This isn't a philosophical claim. It's a result that has been consistently repeated in real experiments.

One of the key questions is: *what exactly counts as observation?* Is it the physical interaction with a detector? Is it the recording of the data? Or does it require a conscious observer — a mind aware of the outcome?

Most physicists explain the observer effect in terms of interaction with a measuring device. When the particle interacts with a detector, it becomes "entangled" with that system, and the wave function collapses. But this raises another question: if the data from the detector is recorded but never seen, is the outcome still determined? Some interpretations suggest that consciousness might be required to "finalize" the result — that a conscious observer completes the process.

This possibility has been both fascinating and controversial. Some scientists argue that bringing consciousness into physics is a mistake — that it introduces too many unknowns and opens the door to speculation. Others point out that ignoring the observer's role creates a gap in understanding, since measurement is not just a physical interaction but a process that connects the system with the outside world — and possibly with awareness itself.

The idea that consciousness might affect matter is not new. In various forms, it has appeared in ancient philosophy, spiritual teachings, and esoteric traditions. But quantum physics gives the idea a different kind of context. It doesn't claim that thought creates reality. It shows that observation changes outcomes — and this change happens even in carefully controlled laboratory conditions.

This doesn't mean you can collapse wave functions just by wishing. The effect is subtle, limited, and still not fully understood. But it suggests that reality at the smallest level is not entirely separate from awareness. The act of looking, measuring, or knowing is not passive. It shapes what is.

Another experiment worth noting is the quantum eraser. In this version of the double-slit test, information about which path the particle took can be "erased" after the particle has already passed through the slits — and the interference pattern returns. Even more strangely, in some setups, the decision to erase the information can happen *after* the particle has hit the screen. This implies that choices made in the present can affect outcomes that seem to be in the past — at least at the quantum level.

Again, the results are not fiction. They've been observed in repeated experiments. What they mean is still debated. But they raise serious questions about time, causality, and the role of information. And at the center of all this is the observer — whatever that word ends up meaning.

Whether consciousness is required for observation or not, the fact remains: measurement changes what's possible. A particle's range of behaviors narrows down into a single result the moment it's observed. That process — from potential to actual — is a defining feature of quantum behavior.

In a system where observation matters, and where measurement collapses probability into fact, the observer is not just watching reality happen. The observer is part of what makes reality happen.

From Particles to Perception

Everything around us is made of particles. The chair you're sitting on, the air you're breathing, the device you're using to read this, your body, your brain — all of it is built from the same basic ingredients: atoms. And those atoms are made of subatomic particles that follow the strange rules of quantum physics. That means the unpredictable, unstable behavior observed in quantum experiments isn't far away or abstract. It's part of what you are.

On the surface, this isn't obvious. People don't vanish and reappear like particles in a lab. We don't experience superposition in our daily lives. The world feels stable, fixed, and mechanical. But that stability may just be the average outcome of countless uncertain events happening at microscopic levels. The randomness smooths out at our scale. That doesn't mean it isn't still there, shaping things behind the scenes.

If the foundation of matter behaves in a way that responds to observation, then it's fair to ask how that principle might relate to perception. Human experience is built from perception. We take in signals — light, sound, touch, temperature — and process them through a system made of particles. Our eyes receive light, our ears capture vibrations, our skin registers pressure and heat. The brain processes all this information and forms a model of the world. That model becomes our reality.

Perception is more than just information, though. It's filtered through attention, shaped by memory, colored by emotion. Two people can see the same event and walk away with different impressions. That's not just psychology — it reflects how deeply personal and subjective perception really is. And if observation shapes quantum systems, then perception may be more than just passive reception. It might play a role in the formation of what appears real.

There's no strong evidence yet that perception, by itself, can collapse quantum states outside the body. But within the system of the observer — meaning within your own sensory and cognitive system — quantum effects may already be part of how experience forms. The brain is made of neurons, which operate through electrochemical signals. These involve the movement of ions and electrons — particles that behave quantum mechanically. That means quantum effects are already embedded in the system that produces thought, awareness, and memory.

Some researchers have proposed that consciousness may arise from quantum processes inside the brain. One theory, called orchestrated objective reduction, suggests that certain structures within neurons — called microtubules — might act as quantum processors. These microtubules could, in theory, sustain quantum coherence long enough to influence brain function at a meaningful level. While this theory remains controversial and unproven, it reflects a growing interest in linking quantum behavior to the workings of the mind.

Even without embracing specific models, it's clear that the distinction between particles and perception is not absolute. The body is built from particles. The senses collect particle-based signals. The brain, made of particles, processes those signals into perceptions. And perception, once formed, directs behavior, emotion, memory, and attention — all of which feed back into how we interact with the physical world.

This loop suggests that the boundary between mind and matter may be thinner than we think. What begins as an interaction at the quantum level eventually becomes a sensation. That sensation becomes part of our internal experience. That experience influences how we move, speak, think, and observe. And those actions feed back into the world, which is still built from the same quantum building blocks.

In this light, perception isn't separate from the physical world. It's part of the same system. And if the system itself is influenced by observation, then our role in shaping the flow of events may be more active than passive. We're not watching from the outside. We're inside the process, made of the same elements that respond to being seen.

Of course, this doesn't mean that by perceiving something, we can force the universe to change direction. It doesn't mean we can think particles into different paths or bend reality with attention alone. But it does raise the possibility that perception is entangled with the structure of reality in ways we don't yet fully understand.

When you observe something, even silently, you're not just registering information. You're taking part in the process that defines what is. And that process — from the movement of particles to the shape of perception — may not be as separate as we once believed.

Chapter 2:
Mind, Matter, and Energy

Thoughts don't weigh anything. You can't hold a memory in your hand or put a belief on a scale. And yet, these things have force. They shape decisions, drive behavior, trigger emotion, and in some cases, affect the body as clearly as a physical impact. The question is: what are they made of?

Everything we call mental — thoughts, attention, imagination, memory — comes from somewhere. And while we often treat the mind as invisible or separate, it works through physical systems. Neurons fire. Signals move. Energy flows. This chapter looks at that flow — how the mind might behave as a form of energy, and what that means for its interaction with the material world.

The Mind as an Energy Field

The mind doesn't exist in a single place. You can't open the brain and point to a section and say, "Here is the mind." What you'll find are structures: cortex, neurons, synapses, pathways. These are physical. They can be measured, scanned, stimulated. But the thoughts themselves — the content — are not visible. They show up only through their effects.

Despite this, the mind produces measurable activity. When neurons fire, they generate small electrical impulses. These impulses don't stay locked in the cells. They move. They create electric and magnetic fields that extend beyond the skull. Sensitive machines, like EEGs and MEGs, can detect them. The brain radiates a field — one that changes depending on what you think and feel.

This isn't metaphor. It's measurable. The brain's electromagnetic field is real, and it fluctuates second by second. Different mental states produce different patterns. Calm thinking has one signature. Stress has another. Deep focus, emotional intensity, or memory recall all change the patterns. The mind, through the brain, generates energy — and that energy moves through space.

Some researchers suggest that this energy field might be more than a byproduct. It could be functional. It might play a role in how different parts of the brain communicate. Instead of neurons passing information purely through chemical signals, the field itself might help coordinate activity across regions. In this view, the mind is not just inside the brain — it also surrounds it, like a field around a magnet.

There are studies showing that heart rhythms and brain activity can synchronize between people during interaction. In close relationships, the nervous systems of two people can influence each other — subtly but measurably. This happens not only through words or expressions, but through proximity alone. Some experiments suggest that when one person focuses their attention on another, the second person's brainwaves show a detectable response, even if they're unaware of being watched.

These effects are small. They're not proof that the mind can reach out and move objects. But they suggest that mental activity might radiate, that it may not be entirely contained. If thought produces a field, and that field extends beyond the body, then the idea that the mind is purely private becomes harder to defend.

The possibility that the mind behaves like an energy field also shifts how we think about memory, attention, and focus. If mental states generate patterns in space — not just in brain tissue — then intention might have a kind of presence. It wouldn't need to be mystical. It would just follow from the physics of fields.

In this light, the mind starts to look like something that flows. Not just in the form of thoughts moving from one neuron to another, but in the way attention moves, how moods ripple outward, how awareness expands and contracts. The mind may behave less like a machine and more like a field system — one that organizes, responds, and interacts.

Some traditions have long described the mind as an energy body, and while those terms aren't used in neuroscience, the basic idea overlaps. Mental states influence the body. They affect the heart rate, breathing, muscle tone, digestion. They alter posture and tone of voice. All these signals feed back into the environment, affecting the people and systems around us. The mind radiates through behavior, through subtle cues, and possibly also through fields.

The strength of this field may be affected by attention. When you focus, you organize mental energy. You reduce noise. You increase coherence. This might explain why deep concentration feels different — not just mentally, but physically. It creates a kind of mental pressure, a force of awareness that seems to stabilize thought and action.

We don't yet know the full extent of how this works. But what's becoming clear is that the mind can't be separated cleanly from matter. It works through the brain, but it also changes the body. It creates signals that move. And those signals can be felt, measured, and in some cases, responded to — even from a distance.

If the mind is a field, then thought is movement. And the way we use that movement may shape more than just internal experience. It may be part of how we interact with the world, whether we realize it or not.

The Brain as a Receiver

The brain is often described as a machine, a processor of inputs and outputs. It takes in signals from the senses, processes them through electrical and chemical activity, and produces thoughts, movements, memories, and emotions. This view has helped drive neuroscience and medical research for decades. But it assumes something that still hasn't been proven: that consciousness is generated entirely inside the skull.

Despite all the progress in brain science, we still don't know how awareness arises. We can trace activity patterns that match certain thoughts or perceptions. We can interrupt specific regions and observe changes in speech, memory, or behavior. But none of that explains why there is a conscious experience attached to that activity. This has led some researchers and thinkers to suggest that the brain may not be the source of consciousness at all. Instead, it might be a receiver.

The idea is not that the brain does nothing, but that it plays a role more like a radio than a generator. A radio doesn't create music — it picks up signals already present in the air and translates them into sound. If the brain works in a similar way, then thoughts, awareness, and even identity may come from something beyond the physical structure itself.

This model would help explain a number of anomalies that don't fit neatly into the standard model. For example, there are cases of individuals who have lived full, functioning lives with much less brain tissue than normal. In rare conditions like hydrocephalus, the brain can be compressed against the skull, yet the person shows normal intelligence and behavior. These cases are rare and not fully understood, but they raise questions about how much of our cognitive capacity depends strictly on brain volume.

If the brain is a receiver, then it would work by tuning in to certain frequencies or fields. This might sound speculative, but it's not without precedent. The brain already operates using electromagnetic signals. Neurons fire through electrical pulses. Those pulses create oscillations — brainwaves — that can be measured from outside the head. These waves shift based on mental states: deep sleep, focused attention, relaxed wakefulness, or stress. The fact that we can detect brainwaves noninvasively shows that the brain emits something – and that it might also be capable of receiving.

Some theories suggest that consciousness may exist as a kind of field – not necessarily in space the way a magnetic field exists, but as a form of non-local information. In this view, the brain would act as an interface, receiving and translating this information into thoughts and sensations. Just like a radio needs the right components to receive certain frequencies, the brain would need specific structures and activity patterns to access awareness.

One early supporter of this view was the psychologist and philosopher William James, who proposed over a century ago that the brain may "transmit" consciousness rather than produce it. Later thinkers, including Aldous Huxley and even some physicists, have echoed this idea in different forms.

From a physical standpoint, this view faces challenges. There is no clear evidence yet of a field or signal that exists independently of the brain. But the fact that brain activity alone does not explain subjective experience leaves open the possibility that something is missing from the current model.

In modern neuroscience, most interpretations of the mind remain strictly material. But cracks are starting to appear in that framework. New research into quantum biology has shown that some biological systems may rely on quantum effects – coherence, entanglement, and tunneling – to function. For example, birds use quantum navigation to detect magnetic fields, and plants use quantum processes for efficient photosynthesis. If these subtle effects operate in biology, it's worth asking whether the brain might also make use of them – especially in the parts related to memory, decision-making, or consciousness itself.

There are even some experiments, still under debate, that suggest focused human intention can produce small changes in physical systems — from random number generators to photon pathways. These effects are weak and hard to replicate, but they raise the question of whether the mind interacts with its environment in ways that go beyond standard sensory input.

If the brain is a receiver, then thoughts are not limited to the head. Mental activity could be something we tap into — filtered and shaped by the brain, but not confined to it. This would mean that awareness is not something the brain creates, but something it connects with. In that case, damage to the brain would not destroy consciousness — it would simply distort the signal.

This model also suggests that consciousness might not be limited to individual minds. If awareness exists beyond the brain, then it might be shared, interconnected, or layered in ways we don't yet understand. That idea would explain why some people report experiences of unity, expanded awareness, or connection with others that feel real — even if they can't be measured.

For now, the brain-as-receiver theory remains speculative. But it addresses questions that the standard model can't answer. It reminds us that while we've learned a lot about the structure and chemistry of the brain, the mystery of consciousness is still wide open — and we may be looking in the wrong place if we assume it starts and ends with neurons.

Emotions, Frequency, and Coherence

Emotions are often treated as psychological events. They come and go, triggered by thoughts, memories, or situations. But they're also physical. When you feel anxious, your heart rate speeds up. When you're calm, your breathing slows. Emotions don't just live in the mind. They move through the entire body. They change posture, muscle tone, digestion, and even the immune system. They're not abstract. They're energy, and they follow patterns.

Modern research in psychophysiology shows that different emotional states produce distinct patterns of activity in the nervous system. The heart, in particular, responds quickly and strongly to emotion. It doesn't just pump blood. It sends signals to the brain — more signals, in fact, than the brain sends to the heart. These signals can either support or disrupt mental clarity, depending on the emotional state.

When you experience a positive emotion — like appreciation, compassion, or gratitude — the heart rhythm becomes smooth and regular. This pattern is known as coherence. In this state, the body's systems come into alignment. Breathing, heart rate, and even brain waves begin to synchronize. This isn't just a subjective feeling of peace. It's a measurable shift in physiological function.

Coherence doesn't mean you're passive or detached. It means your internal systems are operating efficiently. Attention becomes more stable. Reaction times improve. The brain becomes more organized. In this state, it's easier to make decisions, solve problems, and relate to others with less conflict. Coherence is not just emotional balance. It's a form of functional clarity.

Negative emotions — anger, fear, frustration — produce the opposite effect. They create irregular, chaotic heart rhythms. These patterns disrupt the flow of information between the heart and brain. Attention narrows. Thinking becomes reactive. Memory and impulse control weaken. It's not that these emotions are bad in themselves — they're part of being human. But when they dominate, they throw the system out of rhythm.

If emotional states can shift the patterns of our internal systems, then they're also influencing the field the body produces. As mentioned earlier, the brain and heart both generate electromagnetic fields. The heart's field, in particular, is strong enough to be measured several feet from the body. These fields change in real time with emotional state. That means emotions aren't just felt — they're broadcast, in subtle ways, into the space around us.

There are experiments suggesting that people can sense emotional information from others without speaking or even seeing facial expressions. In some cases, physiological data shows synchronization between individuals sitting near each other, especially when they're emotionally connected. Couples, for example, often display mirrored patterns in heart rate and breathing, even in silence. These effects are still being studied, but they suggest that emotional energy has presence beyond thought alone.

The idea of frequency becomes useful here. Every electrical system operates at certain frequencies — rates of vibration. The brain has known frequency bands: delta for deep sleep, theta for dreaming, alpha for relaxed wakefulness, beta for alert thinking, and gamma for high-level integration. Emotions can influence which frequencies dominate. Stress tends to raise beta activity. Calm states promote alpha. Meditation often increases theta and gamma.

These frequencies reflect not just mental activity, but emotional tone. A person in a highly coherent state will often show increased alpha-theta balance, smoother heart rhythms, and better synchrony between brain and body. This isn't mystical — it's biological. But it points to a deeper pattern: emotions are frequency states. They carry signature vibrations, and those signatures affect how the body functions.

Some practitioners and researchers propose that emotional states can be trained like muscles. Just as physical exercise improves coordination, regular emotional self-regulation can improve coherence. Techniques like slow breathing, focused attention on the heart, and memory of emotionally uplifting experiences can help shift the system into a more ordered state. This doesn't erase difficulty or prevent negative emotions from arising, but it helps the system return to balance more quickly.

The broader implication is that emotional energy isn't random. It follows patterns, and those patterns affect both inner experience and outer interaction. A coherent emotional state doesn't just feel better — it changes the body's efficiency, the brain's clarity, and even the energetic environment around a person.

If consciousness has a physical aspect, and if the mind radiates through energy fields, then emotion is one of the most powerful carriers of that energy. It's not just about how you feel. It's about what you broadcast, even without saying a word.

Thought and Intention

Thoughts pass through the mind constantly. Some are clear, others vague. Some arise automatically, others are deliberate. Most are brief, moving from one subject to the next in seconds. But among all these mental events, there's a difference between a thought that drifts and one that's focused. That difference is intention.

Intention is not the same as desire. It's not just wanting something. It's a directed use of attention, with a purpose. When you focus on something — an outcome, a task, a feeling — and hold that focus steadily, you're applying intention. That focus shapes your choices, your energy, and in some cases, the environment around you.

In neuroscience, intention shows up in the brain as organized activity. When a person visualizes a movement, for example, certain areas of the brain light up — sometimes the same areas used for actual movement. Mental rehearsal activates many of the same networks as physical action. Athletes use this to improve performance. Musicians use it to reinforce memory. Intention engages the system as if the outcome were already underway.

But intention may go beyond preparation. Some research suggests that it can affect systems outside the body. Experiments with random number generators, for instance, have shown that the output of these machines can shift slightly when individuals focus their intention on influencing them. The changes are small and controversial, but they have appeared across multiple studies over time, more often than chance alone would predict.

Other experiments involve directed intention toward biological systems — such as the growth of plants, the structure of water crystals, or even human physiological states. Results are mixed, and many are difficult to replicate. But the repeated appearance of slight but measurable changes suggests that intention, when applied with focus and consistency, may influence outcomes beyond the body.

This doesn't mean that intention overrides the laws of physics. The effects, where observed, are subtle and often require large data sets to detect. They're not dramatic or immediate. But they hint at a possible connection between thought and matter — not through force, but through pattern.

One proposed mechanism is informational. If the universe operates partly through fields — gravitational, electromagnetic, or quantum — then perhaps intention acts as a kind of information field. It doesn't push or pull. It organizes. It influences probability. In this view, focused thought adds structure to an otherwise random field of possibilities. It doesn't force an outcome, but it tilts the odds in small ways.

There's also evidence that intention affects the person who holds it. Focused thought sharpens attention. It improves coherence between brain and body. It reduces distraction. People who apply clear intention to their actions tend to have more consistency, more resilience, and more clarity under pressure. Whether or not intention influences the outer world, it clearly shapes the inner one.

This is important, because the line between inner and outer is not always clean. A change in mental state can lead to a change in behavior, which leads to a change in environment. For example, a person who sets a strong intention to remain calm under stress may actually create fewer stressful interactions — not through magic, but through posture, tone, and presence. Intention changes perception, and perception changes response.

It's also worth noting how quickly the mind responds to intention. Attention, once directed, begins to filter information. You start to notice what relates to your focus. This is part of the brain's normal function — the reticular activating system helps prioritize what reaches consciousness. But it also suggests that intention is more than just a background thought. It changes what you see, what you hear, and what you feel.

In many traditions — spiritual, philosophical, or psychological — intention is treated as a creative force. Not because it bends reality, but because it gives shape to potential. Intention turns possibility into direction. It links thought with action.

In modern terms, this means that consciousness may not just observe reality. It might interact with it — through attention, emotion, and mental focus. Thought becomes more than noise. It becomes structure.

There's no need to exaggerate what intention can do. But there's also no reason to dismiss it as irrelevant. Even if its effects are mostly internal, they're still real. And if it does influence the outer world, even slightly, then intention is not just a psychological state — it's a functional part of how reality responds to consciousness.

Chapter 3: The Power of Thought

Thought is the most active force inside the human experience. Before action, there is intention. Before intention, there is thought. It moves quickly, often unnoticed, shaping how we see the world and how we act within it. We rarely question it, but thought is the foundation of nearly every decision we make — and its effects go far beyond internal dialogue.

This chapter looks at what thought does, not only in terms of behavior but also in how it might relate to perception, health, memory, energy, and even the structure of experience itself. We often treat thoughts as private and passive. But they may have more influence than we realize — both inwardly and outwardly.

Do Thoughts Create Reality?

The idea that thoughts create reality is often misunderstood. It's been turned into slogans, simplified into wishful thinking, and wrapped in unrealistic promises. But behind the exaggerations, there's a real question: can what you think affect what actually happens? And if so, how?

Start with the basics. Thought influences perception. What you expect to see shapes what you notice. If you believe someone dislikes you, their neutral expression might feel like rejection. If you expect success, you're more likely to see opportunities rather than threats. This is not magic. It's the filtering system of the brain at work — directing attention, biasing interpretation, filling in gaps.

The brain is not a camera. It doesn't record reality objectively. It constructs a version of reality based on what you already believe, what you focus on, and what you ignore. This means thought doesn't just interpret experience — it frames it. And because behavior follows perception, this framing has real consequences.

A person who believes they're unlucky might hesitate in moments of opportunity. That hesitation becomes self-fulfilling. They miss a chance not because the world was against them, but because they didn't act. On the other hand, someone who holds a mindset of possibility is more likely to try, to persist, to notice openings that others don't. Over time, this creates a pattern — not because the external world was different, but because the internal world shaped how they moved through it.

In this sense, thoughts shape reality indirectly but powerfully. They define how you engage with the world, and that engagement changes outcomes. This alone is enough to say that thoughts matter — not just as feelings, but as active forces in your personal environment.

But some theories go further. They suggest that thought may influence reality more directly. That intention — focused mental energy — might have an effect not only on the person holding it but on physical systems around them. This is where science becomes cautious, and rightly so. It's easy to believe something feels true simply because it aligns with what we want to believe.

Still, there are studies that raise questions. Some involve random event generators — machines that produce unpredictable outcomes, like digital coin flips. In some experiments, subjects were asked to mentally influence the machine's output. The changes were small, but over large data sets, patterns emerged that differed from pure randomness. These effects aren't conclusive, but they suggest that focused thought might interact with physical systems under certain conditions.

Other research has explored the influence of thought on biological systems. Some experiments suggest that people can affect the physiology of another person at a distance — for example, changes in skin conductivity, heart rate, or EEG patterns — when focused attention is applied. These studies are difficult to reproduce consistently, and many scientists remain skeptical. But they keep appearing, and the data continues to challenge the idea that the mind is a closed loop.

There's also the placebo effect — a well-documented phenomenon where belief produces real physical change. A sugar pill, when believed to be medicine, can relieve pain, reduce symptoms, and even alter measurable biological markers. This is not imagined healing. It's a mental state triggering real changes in the body. If thought can affect physiology, it's reasonable to ask where that influence stops.

The reverse is also true. Negative expectations — known as the nocebo effect — can cause harm, even when nothing harmful is present. This shows that thought doesn't need to be positive to be powerful. It simply needs to be believed.

So do thoughts create reality? Not in the sense of controlling weather or bending physical laws. But in the sense of shaping perception, guiding behavior, influencing health, and possibly interacting with probability — yes, they do. The process may not be direct, and the results may not be dramatic, but the effect is real.

Most of reality is not fixed. It's potential — shaped by how we act, how we respond, and what we choose to see. In that space, thought plays a central role. It selects, it filters, it guides. Whether or not it alters matter, it absolutely alters experience — and in the end, experience is what life is made of.

Neuroplasticity and Mental Change

The brain is not fixed. For a long time, it was assumed that after a certain age, brain structure becomes stable and mostly unchangeable. Damage was seen as permanent. Personality was considered set.

Learning new things past childhood was believed to be difficult, if not impossible. These assumptions were wrong.

Research over the past few decades has revealed that the brain is constantly changing. This ability is called neuroplasticity — the capacity of the nervous system to change its structure and function in response to experience, thought, behavior, and emotion. It's not just that the brain can adapt. It does adapt, all the time.

Every experience you have — every movement, every word you read, every new piece of information — leaves a trace in the brain. These traces are physical. They show up as new or strengthened connections between neurons. When something is repeated, those pathways become stronger. When something is ignored, they fade. This is sometimes summarized by the phrase: neurons that fire together, wire together.

This principle is key to how habits form, how memories are built, and how skills are developed. But it also means something more profound: who you are is not locked in. The patterns of thought, behavior, and emotion that seem permanent are in fact flexible. They may be deeply conditioned, but they can be changed.

This change begins with awareness. Many thoughts run unconsciously — automatic reactions, background beliefs, stories we tell ourselves without realizing it. These patterns drive emotion and behavior without conscious approval. But once a pattern is seen, it can be interrupted. That's the entry point for change.

Mental training is possible because of neuroplasticity. Practices like meditation, cognitive behavioral therapy, visualization, and even focused learning reshape the brain. Studies show that consistent mental focus in these areas leads to measurable changes in gray matter density, connectivity between brain regions, and activity patterns in response to stress or emotion.

For example, mindfulness meditation — the practice of paying attention to the present moment without judgment — has been shown to reduce activity in the default mode network, a part of the brain linked with mind-wandering and self-referential thought. At the same time, it strengthens areas associated with attention, empathy, and emotion regulation. These changes are not just subjective. They are visible on brain scans after weeks or months of consistent practice.

Similarly, therapy approaches that target thought patterns — such as identifying and challenging negative beliefs — have been shown to produce structural and functional changes in the brain. The change happens not just through insight, but through repetition. New thoughts become more available. Old ones lose their grip.

This capacity for change also explains why trauma can have such a lasting impact. When a strong emotional experience imprints itself on the brain, it creates pathways that become dominant. The nervous system begins to respond as if the trauma is still happening. But this is also where healing is possible. Because the brain can rewire, it can learn new responses. It can unlearn fear. It can rebuild connection, trust, and self-awareness.

One important part of this process is attention. What you focus on grows stronger. If your attention is constantly directed toward worry, anger, or fear, those pathways become more dominant. If your attention is redirected — even gradually — toward gratitude, stability, and purpose, new connections form. Over time, those connections change the baseline state of mind.

The process isn't quick. The brain resists change, especially when patterns have been reinforced over years. But resistance is not the same as impossibility. Change is a biological fact. The challenge is in sustaining the conditions that allow it to happen.

Mental focus, emotional regulation, repetition, rest, and clear intention are all part of that process. The brain needs time to adjust.

It needs a reason to form new structures. And it needs consistency. That's where most change efforts fail — not in potential, but in follow-through.

There's also a social aspect to neuroplasticity. The environments we move through influence how we think. The people we spend time with reinforce certain neural patterns. Changing mental pathways often requires changing surroundings, conversations, and habits — not just thoughts.

The takeaway is simple: the brain is built to change. It doesn't always want to. It falls into grooves, it prefers the familiar. But it can learn. It can adapt. And most importantly, it can be shaped by conscious effort. That effort begins with what you think, what you focus on, and how consistently you practice it.

Thought is not just a passing moment. It's a tool that rewires the brain, one signal at a time. Neuroplasticity is the mechanism behind that change — proof that growth is not limited to the body, and that transformation is built into the system by design.

Heart-Mind Coherence

The idea that the heart and brain communicate isn't just poetic. It's measurable. While the brain has long been considered the control center of the body, research over the past few decades has revealed something more complex. The heart isn't just a pump. It has its own nervous system — a dense cluster of neurons capable of sensing, processing, and sending information. It communicates with the brain in multiple ways: neurologically, hormonally, electrically, and energetically.

This two-way communication means that the state of the heart influences the state of the brain — not just metaphorically, but directly. And when the signals between them become synchronized, the result is a physiological state known as coherence.

Coherence refers to a pattern of interaction between systems that is efficient, balanced, and stable. In the context of heart-brain interaction, it describes a condition where heart rhythms become smooth and consistent, and this pattern feeds back to the brain, supporting better emotional regulation, focus, memory, and reaction time.

When someone is stressed or anxious, the heart rate becomes erratic. The pattern of beats turns jagged and unpredictable. This irregular rhythm sends signals to the brain that disrupt cognitive function. The brain responds by shifting attention toward threat and away from higher thinking. It becomes harder to focus, make decisions, or access long-term memory.

In contrast, when someone feels appreciation, compassion, or calm, the heart rhythm becomes more stable — a smooth wave-like pattern. This coherent rhythm improves communication with the brain. It supports emotional balance, mental clarity, and physical coordination. This effect can be measured in real time using heart rate variability (HRV), a key indicator of nervous system balance.

Heart rate variability is not about how fast the heart beats, but how much variation there is between beats. Healthy HRV means there's flexibility — the heart speeds up and slows down smoothly, adapting to changes in demand. High coherence tends to show up as a highly ordered HRV pattern, reflecting balance between the sympathetic and parasympathetic nervous systems — the parts of the body responsible for fight-or-flight versus rest-and-recover responses.

What makes this interesting in the context of consciousness is that coherence can be influenced by thought. Focusing attention on the heart, combined with a feeling of appreciation or calm, can shift the body into coherence in under a minute. This isn't a vague mood change — it's a direct shift in physiological rhythm, confirmed by biofeedback devices.

This means that thought — specifically, focused and emotionally balanced thought — can change the body's internal state in real time. And once that state changes, perception, emotion, and behavior tend to follow. This is not wishful thinking. It's a feedback system built into human biology.

Some researchers believe that achieving coherence regularly may improve long-term resilience. People who practice techniques to enter this state show lower cortisol levels, stronger immune response, improved sleep, and better emotional regulation. These aren't mind-over-matter claims. They're examples of how closely the mind and body are linked — and how the heart plays a central role in that connection.

There's also an energetic component to this. The heart's electromagnetic field is the most powerful one generated by the body — much stronger than that of the brain. It extends several feet outward in all directions, and it changes depending on emotional state. Studies have shown that people can sense shifts in others' emotional fields, even without direct interaction. In close relationships, heart rhythms can synchronize, especially during shared emotional experiences.

This supports the idea that coherence is not just an internal phenomenon. It affects relationships, communication, and even group dynamics. A coherent person tends to create more order around them. Their presence is less reactive, more stable. Conversations feel different. Tension drops. It's not about dominance or control. It's about the regulation of inner state — and the fact that inner state radiates outward.

In many traditions, the heart is seen as the center of intuition, emotion, and moral sense. While these ideas have often been treated symbolically, modern science is starting to show that the heart has an actual regulatory role in the body's emotional and cognitive systems.

When the heart and brain work together in sync, the result is greater stability, clarity, and effectiveness.

This doesn't mean coherence is a permanent state. It comes and goes, depending on focus, emotion, environment, and stress. But it can be trained. Practices like heart-focused breathing, gratitude recall, and intentional emotional shifts can help restore coherence when it's lost. And over time, these practices create new neural pathways — reinforcing the connection between thought, emotion, and physiological balance.

Heart-mind coherence is not a mystical concept. It's the measurable synchronization of the body's two most influential systems. And it shows clearly that thought, when focused and emotionally aligned, has the power to shift not only perception, but biology.

The Science of Intention

Intention is one of those words used so often it risks losing meaning. But in a scientific and functional sense, intention can be defined as focused mental energy directed toward a specific outcome or goal. It's more than a passing thought or a wish. It carries direction, consistency, and structure. And in both clinical and experimental settings, intention is starting to be understood as something that can produce measurable effects.

To study intention scientifically is difficult. It's not a material object. It can't be isolated under a microscope or stored in a vial. But it can be observed through behavior, physiological response, and in some cases, statistical deviation from chance. Researchers have explored intention across various disciplines — psychology, neuroscience, medicine, and even physics — each approaching it with different tools, but often arriving at similar conclusions: intention seems to change things.

In medicine, this is most clearly seen in the placebo effect. A person takes an inert substance, believing it to be a treatment, and their symptoms improve. This happens not because of chemistry, but because of belief. The intention to heal triggers measurable changes in the body: endorphins are released, inflammation reduces, and even tissue repair can accelerate. This is not imagination — it's a clear case of the mind influencing biological function.

Intention doesn't need to be conscious to have effect. Expectation, belief, fear — all can act like intentions, guiding the body's reactions without deliberate thought. But conscious intention is different. When someone applies focus deliberately and consistently, results often become more organized and predictable. This has been tested in areas ranging from sports performance to immune response. Athletes who mentally rehearse movements activate the same neural pathways as those who physically train. Patients who visualize recovery often recover faster. The brain treats imagined action much like real action.

Outside of the body, things get more complicated. Some experiments have tried to measure whether intention can influence systems beyond the person holding the thought. One of the most studied tools in this area is the random number generator (RNG). These devices produce strings of numbers based on unpredictable processes — often electronic or radioactive decay. In experiments, individuals are asked to influence the output of these devices using only mental focus.

Over many trials and large data sets, researchers have found small but statistically significant shifts in RNG behavior during periods of focused intention. The deviations from chance are minor, but they're consistent enough to raise serious questions. If verified, this would suggest that consciousness — or at least directed attention — might interact with probabilistic systems in ways we don't yet understand.

Projects like the Global Consciousness Project at Princeton University have taken this idea further, placing RNGs around the world and analyzing their data during major global events. Some of the most striking correlations have occurred during times of intense emotional focus — natural disasters, mass meditations, or international crises. While the data remains controversial and hard to interpret, it points toward the possibility that intention, especially when shared by many people at once, may affect physical systems in subtle but real ways.

The mechanism behind such effects is unknown. Some propose informational fields, others suggest quantum entanglement or coherence fields between human systems and technology. These are all speculative models. But the consistent presence of small anomalies during intention experiments continues to draw scientific interest.

Within neuroscience, intention has been shown to alter brain states. Focused mental imagery changes activity in the prefrontal cortex, motor cortex, and other regions tied to action and self-regulation. Long-term use of intention in meditation or goal setting changes both structural and functional aspects of the brain. These changes aren't abstract. They appear in brain scans. They affect sleep, hormone levels, immune function, and decision-making.

What matters here is not proving that thought can move mountains. It's understanding that mental focus changes the system from within — and possibly, in limited and still mysterious ways, from without. Intention helps organize attention, which in turn influences perception, emotion, and behavior. That feedback loop reshapes experience and alters probability over time.

Intention also interacts with emotion. A clear thought that lacks emotional energy tends to have little impact. But a thought fueled by strong feeling — whether desire, commitment, or compassion — tends to stick. It becomes a pattern. The combination of thought, emotion, and repetition forms the foundation of habits, expectations, and outcomes. Over time, intention leaves a trace — not only in memory, but in behavior, in choices, and potentially in the environment.

Science is still early in understanding how this works. The models are incomplete. But the evidence points toward one clear truth: thought is not passive. Directed thought, held with consistency and clarity, can influence systems in ways that are measurable, even if not yet fully explainable.

Intention, then, is not just a mental event. It's a way of shaping what happens next. Whether through action, energy, or interaction, it changes the course of events — sometimes slightly, sometimes significantly — but always in line with how clearly and consistently it's held.

Chapter 4: The Quantum Self

What you call "you" is usually a list of facts: name, age, history, memories, appearance, preferences. It's a personal profile shaped by your experiences, genetics, choices, and habits. Most of it is physical and psychological. It's visible, traceable, and tied to a body. But what happens when you remove the labels? What's left when memory is quiet, the story pauses, and the body rests?

There's a part of identity that goes deeper than biography – the part that knows it's alive, even when it's not thinking. Some call it awareness, presence, or simply the observer. If quantum behavior is part of the fabric of matter, and if consciousness plays a role in observation, then maybe the self is not just a product of the brain or body. Maybe it's something more distributed, more fundamental, more connected.

This chapter explores that possibility – not as a mystical idea, but as a way of understanding the self beyond the surface.

Physical Identity vs. Quantum Identity

Your physical identity is easy to describe. It's the body you live in, the face in the mirror, the medical records, the voice, the fingerprints. It's the nervous system that reacts to pain, the brain that stores memories, the DNA that shaped your biology. It's practical and real – the part of you that can be seen, measured, and described by others.

This version of self is useful. It allows society to function. It helps organize relationships, responsibility, safety, and survival. But it's also incomplete. It doesn't explain awareness. It doesn't explain the silent experience behind the thoughts, or why you have a first-person perspective at all.

A purely physical model of self breaks down under certain conditions. In dreams, the body is absent, yet experience continues. In deep meditation, thought slows or disappears, yet something remains aware. In trauma, people often report watching themselves from a distance. And in cases of near-death or extreme crisis, some individuals report a loss of identity altogether – yet an intensified sense of being.

These aren't scientific proofs, but they challenge the idea that the self is only the body. There's another level of identity that doesn't depend on physical features, memory, or role. It's the part that experiences directly, without needing to describe.

This is where the idea of a quantum self begins. If the body is made of quantum particles – and those particles behave in ways that include uncertainty, non-locality, and observer dependence – then identity may also have properties that reflect those dynamics. Instead of a fixed, isolated self, there may be something fluid, connected, and responsive at the core of being.

In quantum physics, a particle doesn't have a single identity until it's observed. Before measurement, it exists in a superposition – a blend of possible states. Its behavior depends on the context, the setup of the experiment, and the nature of the observer. That doesn't mean particles are conscious, but it does suggest that identity, at the smallest scale, is not a fixed object – it's a process.

This mirrors human experience more than we tend to admit. You're not the same person in every situation. You shift depending on environment, people, and internal state. When calm, you make different choices than when anxious. When alone, your thoughts are different than when you're being watched. Identity, like quantum behavior, is shaped by observation – both from others and from yourself.

The idea of a quantum identity isn't about claiming that people are particles. It's about recognizing that identity may be a dynamic field, not a static object. The self you think you are may be one version — the collapsed version — selected from a range of potential expressions. Who you become depends on where your attention goes, what you observe within yourself, and how consistent that observation is.

If this is true, then identity is not something you find. It's something you focus. Like a particle collapsing into a definite state, the self you express depends on intention, awareness, and interaction. This doesn't deny biology, memory, or history. It just adds another layer — one that's shaped not by the past alone, but by the present act of attention.

This model also allows for expansion. If the self is a field of possibilities, then change is not a disruption — it's a shift in focus. Growth isn't becoming something new. It's selecting a version of the self that was already available, but unused.

The quantum self is not mystical or abstract. It's the idea that who you are is not finished. It's active. It's unfolding. It includes the body and the mind, but also something more fundamental: the capacity to observe, to focus, and to choose — not only what you do, but how you see yourself in the process.

The Inner Observer

Close your eyes for a moment and watch what happens inside your mind. Thoughts arise — about this sentence, about something unrelated, about yourself, others, the past, the next thing to do. Then a feeling might appear: maybe boredom, maybe curiosity, maybe calm. But in the middle of all that activity, there's something else. The part that notices.

That part is not the thought. It's not the emotion. It's the one watching both. You can call it awareness, presence, consciousness — the name doesn't matter. What matters is that it exists, and you can sense it directly. It doesn't speak. It doesn't react. It doesn't take sides. It simply sees.

In philosophy and psychology, this aspect of experience is often overlooked in favor of behavior, memory, and identity. But it's always present. Even when the mind is full of noise, even when sleep pulls the body into rest, even when language disappears — there's a silent witness behind the activity. This is the inner observer.

The observer doesn't think. It observes thought. It's not the story, but the screen where the story plays. It doesn't take credit or blame, doesn't judge or compare. It doesn't change based on mood or memory. It's there whether you're calm or overwhelmed. Whether you feel strong or lost. Whether you're at peace or in conflict.

One way to understand the inner observer is through attention. Wherever you direct your attention, experience sharpens. When you focus on breath, the body slows down. When you focus on pain, the discomfort increases. When you focus on gratitude, the emotional state begins to shift. Attention is not the same as awareness — but it moves within it.

You can notice attention shifting between the breath, a sound, a thought, a physical sensation. Something inside you is watching all of this — and yet it doesn't move. That something is not a thought. It's not located in a specific part of the body. You can't find it like you would a muscle or a memory. But it's there. It's the condition for experience itself.

In the context of quantum theory, this inner observer becomes even more interesting. Observation, at the smallest scale, is what defines outcome. Before being observed, particles remain in a superposition

— a spread of possible states. After observation, one state becomes real. This isn't metaphor. It's how the system behaves.

If consciousness plays any role in the act of observation — and many interpretations of quantum physics suggest that it might — then the observer inside you is not just passively watching the world. It might be participating in how reality takes shape.

This doesn't mean you're personally controlling atoms or changing matter with a glance. But it does raise the possibility that perception and outcome are linked more closely than we've been taught. The simple act of witnessing — without interference, without judgment — might have more influence than it seems.

In psychological terms, the inner observer is what allows for detachment. When you become aware of a thought without being consumed by it, you gain perspective. That space is the difference between reaction and response. The ability to notice what you feel, rather than becoming what you feel, is what makes self-regulation possible.

This observer is also what allows transformation to begin. You cannot change a pattern unless you see it. You cannot question a belief unless you're aware you're holding it. The observer makes change possible, not because it forces change, but because it makes change visible.

In moments of stress, people often say things like "I lost myself" or "I wasn't thinking." These are signs that attention collapsed into reaction. The observer faded into the background. The more present that inner witness becomes, the less likely we are to be dragged by old patterns. This is one reason why meditation is effective. It doesn't add anything new. It clears space for the observer to be seen more clearly.

The inner observer is also what remains stable through change. Your body ages. Your thoughts evolve. Your roles shift. But something in you has been constant through it all — the part that knows experience is happening. That part doesn't have a name or a history. It doesn't need one. It simply sees.

This idea is not mystical. It's direct and accessible. It requires no belief, only attention. To find the observer, you don't need to achieve anything. You just need to notice what notices. And once that happens, something quiet shifts — a sense of stability, even in the middle of uncertainty.

The observer doesn't solve problems, but it gives space for solutions to appear. It doesn't fight emotion, but it allows emotion to move without becoming trapped. It's not a separate self. It's the ground of experience itself.

To recognize the inner observer is to begin stepping outside of identification — not in a detached way, but in a clear one. It's the first step in understanding that awareness is not something you own. It's something you are.

The Field of Possibility

Most people think of reality as something fixed — a world made of objects, laws, and facts. There's truth in that. You can't walk through walls or breathe underwater without equipment. There are limits. But within those limits, there's far more flexibility than we usually notice. And much of that flexibility starts with how we perceive.

Every choice you make is a fork in the road. Every thought you hold influences what you see, how you act, and how others respond. The possibilities you see — or don't see — shape your behavior. And that behavior changes the path you're on. In this way, possibility isn't just philosophical. It's structural. It shapes the direction of events.

Quantum physics adds another layer to this. At the smallest scale of matter, particles don't exist in fixed states until they're measured. They exist in probabilities — a cloud of potential outcomes. This is called the wave function. It doesn't describe a particle's position or momentum directly. It describes the range of what could happen. When observed, the wave function collapses, and one result becomes real.

This means that potential is built into the structure of matter. What becomes actual depends on interaction — especially observation. If this principle applies even loosely to human experience, then what we call reality may not be as solid as it looks. It may contain a range of possible states — and the self plays a role in which ones take form.

Possibility isn't just out there in the world. It's also inside the mind. Every time you shift your attention, you're selecting a version of experience. Focus on fear, and the world appears dangerous. Focus on a solution, and options begin to emerge. The world hasn't changed — but your interaction with it has. The field of possibilities is the same, but you've chosen a different point of contact.

The brain is not neutral in this process. It acts as a filter. Based on past experience, belief systems, emotional conditioning, and attention patterns, it highlights certain options and hides others. This is efficient — it saves energy and keeps you from being overwhelmed. But it also limits what you perceive.

Neuroscience shows that we receive far more sensory data than we can consciously process. What you notice is a tiny fraction of what's available. This means that your version of reality is always a selection — not a full picture. And the criteria for that selection is shaped by internal states.

When fear is high, the brain becomes more rigid. It narrows options, locks into patterns, and prepares for survival. This is useful in short bursts but harmful over time. Chronic stress locks attention into threat-based thinking, and the field of possibility shrinks.

In contrast, calm and focused attention expands awareness. It increases creativity, allows the brain to access more flexible pathways, and supports insight. People often report that when they're in a relaxed but alert state, new ideas appear. The situation doesn't change — but how they see it does. That shift reveals options they couldn't access before.

Some describe this shift as a jump in perspective. Others describe it as a kind of inner opening. Regardless of language, the result is the same: a wider field of possible responses becomes available. This is not fantasy. It's a measurable shift in cognitive flexibility, emotional regulation, and attentional control.

The field of possibility also exists in relationships, in habits, in the stories we tell ourselves. If you believe that people don't change, you will act in ways that reinforce that belief. If you believe you're stuck, your behavior will reflect limitation. The belief becomes a lens — and what's outside that lens becomes invisible.

But beliefs, like quantum states, can change. And once they do, the range of possible outcomes also changes. Not instantly, not magically — but meaningfully. A shift in internal state creates a shift in external behavior, which creates new feedback, which reinforces the new pattern. This is how a possibility becomes reality.

Some traditions speak of potential as something sacred — a force that lives in silence, in stillness, in attention. From a practical point of view, potential is always available, but rarely accessed. We run through habitual patterns because they're familiar, not because they're true. Breaking those patterns doesn't require effort alone. It requires recognizing that a larger field is already there.

To live with awareness of that field means remembering that your perception isn't the whole picture. It's a slice. And that slice can shift — with presence, with attention, with the willingness to see differently.

The field of possibility isn't about infinite options. It's about real options that were hidden by habit, fear, or bias. It's already here. The question is not whether it exists, but whether you're still enough to notice it — and clear enough to choose.

Awakening the Higher Self

Most of the time, people move through life operating from a familiar version of themselves — shaped by past experiences, social roles, and learned reactions. This version knows how to function, how to plan, how to protect itself. It forms habits, pursues goals, and avoids pain. It's practical, but limited. Beneath it, however, there's something else — a deeper layer of intelligence, perception, and presence that often goes unnoticed. Some refer to this as the higher self.

The higher self is not an alternate personality. It's not a supernatural being. It's the part of consciousness that operates beyond reaction. It's aware but not tangled in thought. It sees but doesn't grasp. It knows without needing to explain. Everyone has access to this aspect of self, but few learn how to listen to it.

One way to recognize the difference between your conditioned self and your higher self is by observing your internal dialogue. The reactive self tends to speak in fear, in control, in comparison. It worries about outcomes, defends old stories, and stays close to the familiar. The higher self speaks in clarity. Not in words, but in direction. It doesn't argue or explain. It waits until you're quiet enough to hear it.

This state isn't produced by effort. It's revealed when noise drops. Often it appears in silence, or in moments when thought slows and presence sharpens — walking alone, watching the sky, waking in the early morning before the mind starts spinning. It's a state of simple awareness, free from judgment, full of space.

You don't become the higher self. You return to it. It's not added to your experience — it's what remains when confusion, fear, and conditioning fade. This doesn't mean abandoning your personality or rejecting your past. It means letting those things be without being ruled by them. It means acting from a deeper place.

In practical terms, awakening the higher self begins with attention. Not scattered attention, but stable, direct noticing. When you pause and observe your reactions without getting caught in them, something opens. That opening is not thought. It's presence. Over time, that presence becomes more familiar than the reactive patterns.

Another aspect of the higher self is intuition. This isn't guessing or emotional impulse. It's a deeper kind of knowing — often quiet, sometimes sudden, but always simple. It doesn't come with justifications. It doesn't compete with logic. It just appears, like a signal. Most people override it with overthinking. But with practice, it becomes possible to recognize and respond to it with more confidence.

The higher self also tends to see connections where the everyday mind sees separation. It doesn't cling to roles or labels. It isn't concerned with how things appear. It focuses on what is. This doesn't make it passive — it makes it precise. It sees clearly without distortion, and from that clarity, action can arise that feels aligned, not forced.

One reason this state is rarely accessed consistently is because the modern mind is trained to prioritize speed, productivity, and external validation. Stillness, depth, and internal clarity are treated as luxuries — or worse, as unproductive. But the cost of that imbalance is disconnection: from self, from others, from clarity.

Practices that support awakening the higher self are simple but often avoided because they require patience and consistency. They include regular silence, self-inquiry, mindfulness, breath awareness, and honest reflection. These don't add knowledge — they strip away noise.

When the higher self becomes more present, the quality of thought changes. It becomes slower, more focused, more rooted. Emotion doesn't disappear, but it no longer dominates. You still feel anger, sadness, joy — but without losing your center. Action becomes cleaner. Speech becomes more direct. Relationships become less tangled in projection.

Importantly, the higher self is not superior to the rest of you. It's not a special state reserved for spiritual practice. It's already here, just usually buried beneath patterns, noise, and defense. The shift is not upward — it's inward. Not into complexity, but into clarity.

Awakening the higher self is not a one-time event. It's a process of remembering, forgetting, and remembering again. Over time, the distance between those moments shortens. And eventually, what began as a quiet presence becomes the place you live from — not all the time, but more often than not.

You don't have to become someone new to awaken this part of yourself. You just need to stop mistaking your surface patterns for the whole story. What you are runs deeper than thought, stronger than mood, and more stable than identity. And that deeper self is always here — waiting, watching, ready when you are.

Chapter 5:
Time, Space, and the Multiverse

We take time and space for granted. They seem to be the most basic features of reality — we wake up in the morning, move through space, age across years, and remember the past while imagining the future. But what if these aren't fixed features of the universe? What if they're frameworks created by perception, not structures that exist independently?

Modern physics, particularly in the quantum and relativistic domains, has shown that time and space are not as rigid as they seem. Time can slow down or speed up depending on speed and gravity. Space can bend. Events that seem simultaneous from one point of view might not be from another. On the quantum level, events can be non-local and even seem to influence each other outside of linear sequence.

This chapter explores the possibility that time and space are not absolutes — and that consciousness may interact with them in ways we don't yet fully understand.

Time as a Mental Construct

Ask someone what time is, and they'll usually point to a clock. Seconds, minutes, hours. A steady progression from past to future. But clocks don't measure time itself — they measure motion. The swing of a pendulum, the movement of a gear, the vibration of an atomic particle. Time, as we experience it, is a perception — not a thing you can touch or hold. It's a structure applied to change.

From a physical standpoint, time is defined by change. One moment becomes distinguishable from another because something has moved, decayed, shifted, or interacted. But if nothing changed — if all motion stopped and no process advanced — could we say time was still passing? Or would it vanish?

Einstein's theory of relativity made it clear that time is not absolute. It's relative to the observer. If you travel close to the speed of light, time slows down for you compared to someone standing still. This isn't science fiction — it's been tested and confirmed. Atomic clocks flown in airplanes tick more slowly than those on the ground. The faster you move, the slower time flows.

Gravity also bends time. The stronger the gravitational field, the more time slows down near it. This has been confirmed with satellites and high-altitude measurements. So already, physics tells us that time is not universal. It's flexible. And what we call "now" is not the same everywhere.

At the quantum level, the idea of time becomes even stranger. Particles can become entangled in ways that make their behavior instantaneously correlated, regardless of distance. This raises questions about causality — how can two events influence each other with no time passing between them?

In some interpretations of quantum physics, events don't happen in a strict order until observed. Until then, they remain in a state of possibility — not past or future, but potential. Time, in this view, becomes something tied to measurement and observation, not something running independently in the background.

From the standpoint of consciousness, this aligns with certain direct experiences. In deep meditation, for example, people often report a loss of time. Minutes feel like hours. Hours pass in what feels like moments. The usual sense of sequence drops away. Something remains — awareness — but without the usual markers of before and after.

Dreams also distort time. A long dream might occur in a matter of seconds. A dream that seems to take place over days or weeks may occur in a flash of REM activity. The mind, unanchored from the body and external signals, generates its own sequence — one that often doesn't match physical time.

Memory further complicates the idea. The past isn't real in any physical sense. It exists only as neural patterns and emotional traces. The future is pure construction. The only place either of them exists is in the mind. Yet we live inside both constantly, revisiting past events and imagining what's ahead. This constant mental time travel can be useful — but it also creates anxiety, regret, and distraction.

What's often missed is the fact that experience only ever happens now. The future is never experienced as "the future." When it arrives, it becomes the present. The past can't be touched — only reconstructed. All experience is current. And yet the mind continues projecting forward and backward, building a timeline that feels solid but exists only in awareness.

This doesn't mean time doesn't exist at all. It means time, as we live it, is created by perception. The brain constructs a sequence to make sense of experience. It builds a timeline from memory and anticipation. But this timeline is not an objective feature of the universe. It's a mental model — one that can change based on attention, emotion, and even belief.

When people shift into states of flow, presence, or heightened awareness, time often fades. The mind stops tracking past and future, and attention settles fully into the moment. In these states, action becomes smoother, perception becomes clearer, and the sense of being separate from what's happening begins to dissolve. The clock may still tick, but it no longer feels like a cage.

Understanding time as a mental construct doesn't mean ignoring schedules or abandoning plans. It means recognizing that the pressure of time — the stress, the urgency, the sense of being behind — is often a psychological creation. Events happen. Deadlines exist. But how we experience them is shaped by the lens we use. And that lens can change.

By shifting focus from what was or what might be to what is, the mental weight of time begins to lift. Experience becomes more direct, less filtered by narrative. And in that space, something deeper becomes available — a kind of clarity that doesn't depend on the clock.

Parallel Realities and Quantum Choice

Most people live as if the path they're on is the only one. Every choice made seems to cancel out all the others. Once something happens, it feels permanent. A word spoken, a move made, a decision taken — all locked in. But quantum theory suggests a very different model of reality. One where choice might not collapse possibilities, but split them. One where reality may not be a single track, but a set of overlapping timelines.

This idea is known as the many-worlds interpretation of quantum mechanics. It's not science fiction. It's a serious — though still debated — framework proposed in 1957 by physicist Hugh Everett. According to this model, when a quantum event occurs, all possible outcomes continue to exist, each in its own version of the universe. Instead of one outcome becoming real and the others disappearing, the universe splits. Every possibility plays out — somewhere.

This interpretation avoids the need for a "wave function collapse," the moment in standard quantum mechanics where the field of possibilities narrows to a single result upon measurement. Instead, it claims that all outcomes are real, just not all in the same world.

If this is true — or even partially true — it has implications far beyond physics labs. It means that reality might be branching constantly. Every decision you make would move you into a new version of experience, one among many, with the others continuing along their own paths, outside your awareness.

From the perspective of everyday life, this may seem abstract or impossible to verify. But even at the psychological level, the idea of multiple versions of self, or parallel potentials, makes sense. Every time you choose one action over another, you're not just picking a direction — you're reinforcing a pattern, a version of who you are. The person you become is shaped by these micro-decisions, which add up over time.

Quantum choice, in this broader sense, isn't just about particles. It's about the structure of possibility. If we think of the self not as a fixed entity but as a shifting pattern — a field of behaviors, thoughts, and attention — then each choice you make is a selection. Not just of outcome, but of identity. The person who reacts with anger is different from the one who responds with patience. The one who avoids change lives a different version of life than the one who embraces it.

In this way, parallel realities aren't just out there in the fabric of space. They exist inside potential — in the gap between thought and action, between intention and behavior. The brain is capable of imagining multiple futures, rehearsing possible outcomes, and then choosing one. That process reflects the same kind of branching logic found in quantum systems.

Some researchers in consciousness studies have speculated that the mind might access these possibilities in subtle ways. In dreams, altered states, or deep creative flow, people often describe events or ideas that feel like glimpses into other versions of life — not memories, not fantasies, but something in between. Whether these are true connections or symbolic constructions is still unknown, but the experience of parallel versions of self is widespread across cultures and histories.

Quantum decision-making also implies a responsibility. If multiple versions of reality are possible, then the way you act now doesn't just affect the immediate situation — it defines which path you continue on. Even small, consistent shifts in focus, attitude, or action can gradually move you into a life that looks and feels different from the one you started in. Change doesn't require a dramatic event. It just requires repetition in a different direction.

This concept also supports the practice of visualization and intention-setting. When you imagine a different version of yourself — more focused, more open, more creative — you're rehearsing a shift in trajectory. The more clearly you imagine it, and the more your actions align with it, the more likely you are to experience a version of reality where that version becomes your default.

That's not magic. It's feedback. Each choice changes the conditions for the next one. And that chain of change builds a new path — just as surely as a particle moving through space leaves a trail.

If reality contains layers of potential, and if consciousness helps select among them, then the world is not fixed. It's responsive. And choice is not just about results. It's about defining the version of life — and the version of self — that becomes real.

The most powerful part of this isn't theoretical. It's practical. You're already choosing. The question is whether you're choosing by habit, by fear, by conditioning — or by attention, by clarity, and by intention. Every moment is a fork. And the path you're on is not the only one that exists.

Synchronicity and Subtle Order

Sometimes, events line up in ways that feel too precise to be random. You think of someone you haven't seen in years, and they contact you the same day. You're struggling with a decision, and a conversation overheard in passing gives you the exact words you needed. There's no clear cause, no chain of events that explains it. But something about the timing feels meaningful. That's synchronicity.

The term was coined by Carl Jung, who described synchronicity as the meaningful coincidence of events with no apparent causal link. Unlike random chance, synchronicity carries a subjective weight — a feeling that something aligns beyond logic. Not just coincidence, but connection.

From a scientific perspective, these events are difficult to study. They're not predictable. They don't follow repeatable patterns. But they show up in personal narratives across cultures, often during moments of emotional intensity, transition, or inner conflict. People don't just notice them — they're often changed by them.

Synchronicity points to a different kind of order. Not the mechanical order of cause and effect, but a subtle pattern that connects internal states with external events. It suggests that consciousness and environment might reflect each other more closely than we assume.

This doesn't mean events are scripted, or that everything happens "for a reason" in the traditional sense. It means that patterns may emerge based on how we direct attention, energy, and emotion. When the mind is quiet, present, and receptive, it's more likely to notice connections. When attention is scattered, lost in mental noise, those same patterns pass by unnoticed.

Quantum theory doesn't explain synchronicity directly, but it does leave room for a more flexible view of causality. In classical physics, A causes B. In quantum mechanics, outcomes can be influenced by conditions that don't behave in linear time. Entangled particles respond to each other instantly, regardless of distance. Choices made in the present seem to influence how past measurements unfold. This doesn't translate neatly into everyday events, but it does show that the flow of cause and effect may be more layered than we think.

Synchronicity might be one way that this layered causality expresses itself in lived experience. Not through magical thinking, but through resonance — the alignment of inner and outer states. When your internal world reaches a certain clarity, the external world seems to respond. Not because you caused something, but because you're tuned into a pattern you hadn't seen before.

This tuning doesn't require belief. In fact, belief can sometimes interfere — creating expectations that force patterns where none exist. True synchronicity doesn't come from trying to control events. It appears when attention is open, not when it's grasping. That openness is what allows meaning to surface in ordinary moments.

From a psychological point of view, synchronicity reflects the brain's ability to detect patterns. When something emotionally important is unresolved, the mind scans for related signals. It becomes more sensitive to information that fits the inner state. This is part of how intuition works. It's also part of how meaning is made.

But there's something more than mental pattern recognition at work. Many synchronicities happen with no prior emotional buildup — they arrive unexpectedly, quietly, but with a precision that goes beyond what the mind was looking for. They carry a quality that's hard to describe but immediately recognized. They stop thought for a moment. Something lands.

People often report more synchronicity when they're in a period of inner transition. During grief, change, healing, or creative focus. These are times when the usual mental filters loosen, and attention becomes more fluid. What was ignored before suddenly becomes visible. Small events begin to feel significant — not because they're louder, but because you're finally listening.

The key to synchronicity may be presence. Not hypervigilance, not control — but quiet, grounded awareness. In that state, experience becomes richer, and connections that once seemed invisible begin to emerge. Life feels less like a string of separate events, and more like a woven whole.

This doesn't mean every coincidence is meaningful. Randomness exists. But not everything that looks random is empty. The challenge is to stay clear enough to sense when something touches that deeper layer — not to force meaning, but to recognize it when it comes.

Synchronicity reminds us that the world may be more responsive than we've been taught. Not always predictable, not always logical, but full of quiet signals — waiting for attention to become still enough to notice. And when that stillness appears, the line between inner and outer starts to blur. The world becomes more like a mirror — not reflecting what we want, but what we're ready to see.

Quantum Shifts in Consciousness

Most change happens slowly. Habits form, thoughts repeat, identity solidifies over time. But once in a while, something shifts suddenly. A way of seeing collapses. A belief dissolves. A tension that felt permanent disappears in a moment. It's not always clear what triggered the shift — a conversation, a silence, a moment of clarity — but something in the mind rearranges itself, and you're no longer in the same place.

These moments are often called breakthroughs, awakenings, or turning points. But they don't always come with drama. Sometimes they arrive quietly, like a window opening. You see the same world, but differently. The structure of experience changes, even if the contents don't.

These are quantum shifts in consciousness. The term doesn't mean that the mind becomes subatomic, or that consciousness literally jumps across particles. It's used here to describe the nature of certain internal changes that resemble what happens at the quantum level:

sudden reorganization, non-linear transitions, unpredictable but transformative results.

In classical systems, change tends to be gradual and proportional. Small inputs lead to small effects. In quantum systems, this isn't always the case. A particle can remain in a stable state until just the right condition appears, and then it jumps — discontinuously — to a new state. These transitions can't be predicted precisely. They don't follow the normal path of cause and effect. They seem to unfold from a set of possibilities, with the actual result appearing only at the moment of interaction.

Consciousness sometimes behaves the same way. A person can repeat a pattern for years, fully aware of it but unable to shift. Then something changes — and the whole pattern drops. Not because it was solved piece by piece, but because the structure that held it together lost energy. The system reorganized. And in its place, something new appeared.

This kind of shift often happens when awareness deepens without resistance. Not through force, but through openness. When you stop trying to fix or control the mind, and instead observe it clearly, space is created. And somctimes, in that space, a transformation occurs that couldn't be forced from the surface.

It's tempting to think these shifts are caused by specific techniques. Meditation, therapy, breathwork, altered states — all of these can help. But the shift itself rarely comes from the method alone. It comes from a readiness in the system. A loosening of the old framework. A moment where possibility outweighs inertia.

You can't schedule these shifts. But you can create the conditions for them. Clarity, honesty, stillness, and attention increase the chances that change won't be blocked by habit. The deeper you listen, the more likely something will rearrange itself when the time is right. It's not a matter of effort, but of alignment.

One of the most disorienting parts of a quantum shift in consciousness is that it often lacks explanation. You just know something is different. A fear you carried is gone. A reaction you once believed in doesn't activate anymore. You feel lighter, more neutral, or more aware — without having done anything dramatic to achieve it.

Sometimes the shift is temporary. The old patterns try to return. But even then, the memory of that new space remains. You know what clarity felt like. And that memory becomes a reference point. It shows you what's possible — not as an idea, but as something you experienced directly.

Quantum shifts are not constant. Most of life moves in steps, not leaps. But when these leaps occur, they change your baseline. You don't go back to being the same. Even if the world looks the same, your relationship to it changes. You start acting from a different place. You make different choices. You see different outcomes.

This kind of transformation doesn't mean becoming perfect or transcendent. It means becoming more accurate — more in touch with what's real, less caught in the loops that used to define you. It's not enlightenment. It's recalibration.

And often, the moment of shift comes not when you're trying hardest, but when you've finally stopped trying. When the noise settles. When control loosens. When attention rests fully in the present. That's when the system has room to change — not gradually, but fundamentally.

These shifts remind us that consciousness isn't a fixed state. It's alive, layered, and full of thresholds we haven't crossed yet. The mind isn't a closed loop. It's a field of potential, waiting for the right moment to take a new shape.

Chapter 6: Intuition, Telepathy, and Mental Connection

Not all forms of knowledge come from logic. Some appear suddenly, without explanation, as a sense, a pull, or a clear internal "yes" or "no." These moments often bypass analysis. They don't follow a path of reasoning. Yet they guide decisions, predict events, or reveal insights that weren't reached through thought. We call this intuition.

Other experiences stretch the limits of the separate mind even further – sensing someone's emotional state without words, finishing a person's sentence before they speak, or feeling someone's presence before seeing them. These moments are subtle, hard to study, and often dismissed as coincidence. But they persist.

This chapter explores these experiences – not to prove or disprove them, but to understand what they suggest about the nature of thought, perception, and connection. If the mind is not just isolated activity within the skull, but a field that interacts with others and the world, then phenomena like intuition and telepathy may point to a level of intelligence we're only beginning to recognize.

Intuition as a Non-Linear Form of Knowing

Intuition is often described as a gut feeling, a sense of knowing without knowing how. It doesn't arrive as a complete argument or detailed explanation. It appears as a direction – subtle but immediate. The challenge is that it doesn't wait for confirmation. It doesn't present evidence. It just shows up.

In a world built on logic and proof, intuition is often treated as unreliable. But in practice, it plays a central role in decision-making – especially in situations where time is short, data is incomplete, or complexity is too high to process analytically. Intuition can guide action before thought catches up. And often, it turns out to be right.

This kind of knowing is not irrational. It's non-linear. Instead of building step by step, like traditional reasoning, intuition draws from patterns too complex to trace consciously. The brain processes far more information than we're aware of — from body language, tone of voice, context, memory, and previous experience. It integrates this data in the background, and when something aligns or misaligns, the result surfaces as a feeling.

But intuition isn't just subconscious pattern recognition. In some cases, it seems to go beyond stored experience. People describe knowing things they couldn't possibly have known — a danger before it appears, a truth they hadn't been told, a sense about a person they've just met. These moments don't come from the past. They feel immediate, like perception, not memory.

Some researchers in consciousness studies suggest that intuition may be a direct interaction with information fields not limited to the brain. If thought exists in fields — as electromagnetic, quantum, or even non-local structures — then perception might not be confined to sensory input. Intuition may be the result of tuning into signals that are real but subtle — signals the conscious mind can't decode, but the deeper system can sense.

This tuning process may be linked to coherence. As seen in earlier chapters, when the mind and body are in a coherent state — especially through regulated emotion and focused attention — the system becomes more sensitive. Patterns become easier to detect. Noise drops. Signal becomes clearer. Intuition often arises in these moments — not in confusion, but in clarity.

Intuition also strengthens with practice. People who learn to listen to it, test it, and refine their sensitivity often report that it becomes more reliable over time. It's not about guessing. It's about learning the difference between noise (fear, projection, fantasy) and signal — that clean, quiet sense that something is simply true.

One of the main barriers to intuition is mental noise. When the mind is filled with worry, calculation, or repetition, the signal gets buried. Intuition doesn't shout. It waits. When space opens, it speaks. This is why intuition often arises when the mind is doing something else — walking, showering, just waking up — because the conscious effort to "figure it out" has relaxed, and the deeper intelligence has room to speak.

This kind of knowing doesn't replace analysis. It complements it. In fact, the best decisions often combine the two: intuition opens the direction, logic helps structure the action. The two systems can work together — if we trust each for what it does best.

In creative work, intuition plays a central role. Artists, musicians, writers — all speak of inspiration as something they don't generate but receive. It doesn't come from analysis. It arrives suddenly, often whole. The same happens in science, where moments of insight often precede the step-by-step verification. Something is seen before it is proven.

The deeper question is: where does this knowing come from? Is it the product of deep, unconscious computation? Is it access to a shared mental field? Is it a memory from a version of self that exists outside time? Each answer opens a door to how consciousness might be structured — not as isolated activity, but as part of a wider intelligence.

Regardless of the source, intuition remains one of the most direct, personal proofs that not all knowing follows a linear path. And learning to recognize it — not blindly, not impulsively, but clearly — may be one of the most valuable skills in navigating both the inner and outer worlds.

Telepathy: Perception Beyond Language

We're taught that communication happens through sound, gesture, writing — systems that rely on physical signals passing from one person to another. But many people have experienced moments when no signal was needed. You think of someone, and they call. You feel a mood shift in the room before anyone speaks. You say what another person was about to say. These are often brushed off as coincidence. But what if they're not?

Telepathy is the idea that the mind can access or transmit information without using the usual sensory pathways. It doesn't involve sound, sight, or touch. It happens silently, instantly, and often without any clear mechanism. While mainstream science remains skeptical — due to the difficulty of replicating results — the phenomenon refuses to disappear.

Across cultures and histories, reports of telepathic experience are widespread. Twins who sense each other's pain from across continents. Parents who feel something shift the moment their child is in danger. Close friends who share thoughts without speaking. These stories are not rare — they're just hard to measure.

From a scientific perspective, the challenge is clear: telepathy can't be turned on and off like a light switch. It doesn't respond well to testing under pressure. But there have been controlled experiments — some with surprising results. In certain studies, participants placed in separate rooms have shown correlated brain activity when one is exposed to emotional or sensory stimuli. In other tests, people have been able to identify images, emotions, or mental impressions with accuracy above chance, even with no physical connection between them.

These results are inconsistent, and critics point to the need for better controls. But the data continues to suggest that something is happening — something that can't be fully explained by coincidence, expectation, or normal sensory input.

One possibility is that minds are not as isolated as they seem. If consciousness is not confined to the brain — if it exists as a field or network — then mental connection would not require traditional communication channels. Thoughts would not need to travel through space. They would emerge in shared space.

Quantum theory offers some potential models. In entanglement, two particles that have interacted remain correlated, no matter how far apart they are. Change the state of one, and the other responds instantly. There's no signal passing between them — just a connection that exists outside time and space. If minds can form similar bonds — through emotion, attention, or shared experience — then information could appear in more than one place at once.

Another possible explanation involves coherence. When two systems become synchronized, even weak signals can pass between them more easily. Human brains, especially when emotionally attuned, may generate fields that align. In states of calm, trust, or deep presence, these fields may become more receptive. This might explain why telepathic experiences often happen between people who are emotionally close or in altered states — sleep, meditation, grief, or intense focus.

Still, most people who report telepathy don't claim to have control over it. It happens spontaneously. Sometimes it's helpful. Sometimes it's unsettling. But it's rarely forced. It arises when the usual noise drops — when thought slows, when connection deepens, when attention becomes still.

What's important is not proving telepathy in a lab, but recognizing what it reveals about mind and relationship. It suggests that the boundaries we take for granted — between thoughts, between people — might be more flexible than we think. That connection may not require language. That empathy may be more than emotional — it might be perceptual.

Language is useful, but it's slow. It compresses meaning into words, which the other person must unpack and interpret. Telepathy, if it exists, would bypass that. It wouldn't send sentences. It would transmit states, images, impressions — whole patterns of experience. Not perfectly, not reliably, but directly.

This kind of communication, even in its most subtle form, could explain the sense of deep understanding people sometimes feel without speaking. A kind of resonance. You're not hearing thoughts — you're sensing presence, intention, emotion. Something in you aligns with something in them. No explanation needed.

It's likely that whatever telepathy is — if it's real — it operates beneath conscious control. Like intuition, it arises in stillness. Like emotion, it signals more than words can say. And like attention, it can be trained, refined, or ignored.

You don't need to believe in telepathy to notice its traces. Start with presence. Notice what you pick up from others without them telling you. Observe how much you can sense before a word is said. And consider the possibility that thought might not be as private as it feels. Not in a way that threatens — but in a way that connects.

The Mental Web: Shared Fields of Thought

Thought feels like a private event. You think something, and only you know it. Your mind is locked inside your skull. But that assumption may be more a habit of perspective than a proven fact. In daily life, people often experience something different — especially in close relationships, crowds, or moments of deep emotional intensity. Ideas seem to move between people. Moods spread without words. A phrase forms in your mind, only to be spoken by someone else seconds later.

These aren't just occasional curiosities. They point toward a possible structure behind experience — a shared mental layer that connects individuals not just emotionally, but cognitively. Call it a collective field, a mental web, or group consciousness. The idea is the same: that thoughts, feelings, and mental states don't only exist in isolation, but can be transmitted, shared, or even co-created.

In science, this idea has no established framework — but there are analogies. In nature, systems exist that behave in coordinated ways without a central command. Birds fly in formation, fish turn simultaneously, and flocks of insects move as one organism. The pattern emerges from the interaction of many individuals responding to subtle signals.

The brain does something similar. Neurons fire based on both internal and external inputs, creating patterns that give rise to thought. If each brain is a complex network, then it's possible that groups of brains interacting closely could form temporary networks of their own — not through language, but through shared states.

There's already evidence that this happens in basic forms. In group meditation studies, synchronized heart rate variability and brainwave activity have been observed, even in silence. People sitting together with a shared intention or focus tend to enter similar physiological rhythms. Their nervous systems begin to mirror each other. This process, called entrainment, suggests that mental and emotional states can influence each other in ways that go beyond visible behavior.

But it goes further. In emotionally charged situations — a concert, a protest, a religious gathering — people often report feeling a collective energy, something larger than individual emotion. It's not just agreement. It's a kind of fusion, where individual awareness blends into a shared experience. That field often has its own momentum. It carries a mood, a rhythm, and sometimes even a sense of meaning that's hard to put into words.

The mental web may also explain how innovation moves. In many cases, similar discoveries or inventions appear in different places at the same time, without direct communication. This has happened repeatedly in science, art, and technology. Some call it parallel discovery. Others suggest that certain ideas emerge when the collective field is ready — not from one person, but through many minds tuned to similar frequencies.

This model doesn't require abandoning individual mind. It just reframes it as permeable — not a closed system, but a participant in a larger one. In this view, your thoughts are not only yours. They're shaped by the mental environments you inhabit: the people you spend time with, the culture you engage with, the stories you absorb.

You're both a transmitter and a receiver. When your state shifts — toward clarity, tension, calm, or confusion — you affect others, often without noticing. And when you're immersed in a group, your internal state reflects more than just your personal experience. You absorb, reflect, and amplify what's around you.

Understanding this changes responsibility. It means that mental hygiene isn't just personal. The way you manage your focus, attention, and emotional state influences more than your own mind. It affects others — subtly, quietly, but consistently. It also means that cultivating clarity in your own system may be one of the most impactful things you can do for those around you.

Some traditions have long held this view. Certain forms of prayer, meditation, or ritual are done not only for individual benefit, but for collective balance. The idea is that mental coherence spreads. It doesn't push or preach. It stabilizes the field — like a tuning fork that brings other instruments into harmony without effort.

The mental web isn't a mystical concept. It's an observational one. You can test it in daily life. Notice how moods shift in a room when someone calm walks in. Observe how ideas travel faster in groups that trust each other. Pay attention to how silence, presence, and intention influence conversation more than words alone.

If the mind is not just a container but a participant in shared fields, then attention becomes more powerful than we assume. It doesn't just shape your experience. It becomes part of a wider pattern. The more aware you are of that pattern, the more deliberately — and quietly — you can influence it.

The Ethics of Mental Influence

If thought can influence others — not just through words or actions, but through presence, intention, or subtle connection — then the question is no longer whether it happens, but how to relate to it responsibly. Influence is part of every interaction. We affect each other constantly, often without realizing it. But what happens when that influence becomes intentional? What are the boundaries?

Mental influence doesn't have to be manipulative to be real. A glance can shift someone's state. A word spoken with precision can redirect a thought pattern. Silence itself can become a form of communication. These effects aren't magical — they're human. But they become more sensitive, more significant, when we accept that attention carries weight.

In some contexts, influence is direct and visible — persuasion, teaching, leadership. In others, it's invisible — emotional tone, energetic presence, focused attention. When someone speaks from clarity, others feel it. When someone enters a room full of tension, that tension spreads. Even without physical cues, mental states ripple outward. And once we recognize this, the question becomes: what are we doing with that power?

Intent matters. Are we trying to control others' thoughts or open space for them to think freely? Are we using attention to direct, or to dominate? The line can be subtle. Influence isn't always harmful, but it becomes dangerous when it's tied to unconscious needs – the need to be right, to be followed, to be admired, to be in control.

Mental influence becomes ethical when it's grounded in awareness. That means paying attention not just to what we're doing, but to why. It means watching for attachment – to outcomes, to agreement, to authority. It means allowing others their own timing, their own process, even when we believe we know what's best.

This is especially important in close relationships. The more emotionally connected we are to someone, the more influence we have – and the more we're influenced in return. This creates a feedback loop. If one person moves into fear, the other feels it. If one moves into presence, the whole system can stabilize. This doesn't mean we're responsible for others' states, but it does mean our presence matters.

One of the quietest forms of mental influence is expectation. We don't just respond to others – we shape their behavior by how we see them. If you expect someone to fail, you'll unconsciously reinforce that view. If you hold someone in a state of possibility, they may rise to meet it. This isn't about pretending or projecting positivity. It's about where you place your attention – on the version of them that's stuck, or the version that's growing.

This also applies to self-talk. The way you influence yourself mentally – the patterns you repeat, the identities you reinforce – shapes how you show up in the world. And that presence influences others. Personal clarity becomes collective clarity, not through preaching, but through alignment.

Ethical influence is not about having no effect. It's about being transparent with yourself. It's about letting go of hidden agendas. It's about respecting boundaries, both verbal and nonverbal. In subtle mental spaces, consent still applies — not in the legal sense, but in the relational one. You can feel when someone is pushing energetically, even without words. You can feel when someone is holding space rather than trying to shape you.

In this way, ethics begins in stillness. The more still your own system is, the less likely you are to impose on others. Influence remains, but it becomes clean. Not neutral — but honest. When you speak, act, or even simply observe, you do so from clarity, not need.

There's also humility in recognizing that influence is mutual. You're not the only one shaping the field. You're part of a shared environment. What you offer matters, but so does how you receive. Listening is as powerful as speaking. Attention is as shaping as intention.

As sensitivity increases — whether through personal growth, mindfulness, or deeper states of awareness — the ethical dimension becomes harder to ignore. You start noticing when you're pushing too hard, when you're projecting onto others, when you're using insight to position yourself rather than to understand.

But with that sensitivity comes the opportunity for a different kind of interaction — one rooted in presence, trust, and quiet respect. A space where thought doesn't dominate, but clarifies. Where influence isn't forced, but offered. Where connection doesn't mean control, but mutual recognition.

In that space, communication becomes more than words. And thought becomes something more than personal. It becomes part of a shared field — one that grows clearer, deeper, and more grounded the more carefully we meet it.

Chapter 7: Emotion, Energy, and Frequency

Emotion is often treated as a reaction – something that happens in response to events, thoughts, or memories. But emotion is also movement. It shapes perception, changes physiology, affects decisions, and influences the space around us. More than reaction, it's a form of energy – subtle, immediate, and powerful.

Every emotion carries a tone. It has a pace, a rhythm, a weight. Some contract the body; others expand it. Some sharpen thought; others dull it. We usually name emotions in categories – fear, anger, joy, grief – but their effects go deeper than labels. They condition how attention moves, how we relate, and how we show up in space.

This chapter looks at emotion as energetic frequency – not as metaphor, but as a real aspect of experience. If we understand emotional states as vibrational fields, we begin to see how much influence they have on the quality of mind, health, and interaction.

Emotional Energy and Vibrational States

Emotions are not just mental events. They manifest physically: in breath, posture, muscle tone, heart rate, voice. When you're afraid, the body contracts. When you're at ease, it opens. These shifts aren't symbolic. They're energetic. Emotion changes the internal environment in ways that can be measured – and felt.

At the most basic level, emotional states are patterns of energy in motion. The word itself – e-motion – points to this. Emotions move through the body, often faster than thought. You feel something before you can explain it. That feeling isn't just information. It's frequency.

Each emotion seems to have its own range of frequency — not in the sense of radio waves, but as a quality of internal rhythm. Anger tends to feel hot, sharp, fast. Sadness is slower, heavier, denser. Joy moves light and quick. Calm has its own pace — steady, smooth, even. These patterns affect not only your inner state, but also your capacity to think, speak, and act clearly.

When the system is in a coherent emotional state — like calm focus or genuine appreciation — physiological systems align. Heart rate variability increases, nervous system balance improves, and brain function becomes more integrated. This is measurable. Tools like EEGs and heart coherence monitors confirm it. The body functions more effectively when emotional energy is balanced.

On the other hand, when emotion becomes dysregulated — through stress, repression, or overload — the system becomes noisy. Thought becomes reactive. The body shifts into defense. Clarity drops. Attention fragments. The quality of experience contracts.

This isn't a moral judgment. All emotions serve a function. Fear protects. Anger mobilizes. Grief integrates. But problems arise when emotional energy becomes stuck — when it loops without resolution, or when it's suppressed below awareness. In those cases, the energy doesn't disappear. It stays active, shaping perception and behavior from the background.

What we call mood is often the residue of unprocessed emotion. You walk into a space and feel off, even if nothing's wrong. Or you wake up low without a clear reason. These states are not random. They're signals — not always obvious, but always traceable. And learning to track emotional energy, like tracking wind or water, gives you access to a deeper layer of intelligence.

Some traditions have long spoken of vibrational states. They describe emotion not just in terms of psychology, but frequency — saying that fear contracts the field, love expands it, gratitude lifts it, shame lowers it. These are not mystical beliefs. They reflect a practical observation: certain states generate clarity, others cloud it.

When you're in a low-vibration state, perception narrows. You see fewer options. You anticipate threat. You're more likely to repeat old patterns. In high-vibration states, awareness widens. You access creativity, patience, insight. The conditions for change become more available.

Importantly, "higher" doesn't mean better in a superficial sense. It means more integrated — states where thought, feeling, and action begin to align. This is not about chasing pleasant emotions. It's about tuning into states that support stability and openness.

Regulating emotional energy doesn't mean suppressing anything. It means becoming aware of it. Naming it. Allowing it to move. Emotion, when felt clearly, often shifts on its own. Suppression locks it in place. Awareness lets it breathe.

Breath is one of the most direct tools for working with emotional energy. Slow, even breathing changes heart rhythm, which affects brain state. Movement helps too — walking, shaking, stretching — especially when done with attention. These don't solve emotion, but they help the system process it instead of storing it.

Also important is language. The way you describe your state influences your relationship to it. Saying "I am angry" identifies with the emotion. Saying "Anger is present" creates space. This small shift changes how energy moves. It moves you from reaction into witnessing. And from there, choice becomes possible.

If emotional states are forms of frequency, then managing your emotional energy isn't about control — it's about tuning. Like adjusting the dial of a radio, you don't invent the station. You just choose where to listen. And that choice shapes what becomes available — in thought, in perception, in action.

Your emotional field also affects others. You don't have to say anything. People can feel agitation, ease, tension, openness. These signals aren't imagined — they're picked up through micro-expressions, tone, and nervous system feedback. This is why working with your own emotional energy is not just personal. It's relational.

Learning to feel emotional frequency clearly — without story, without avoidance — is one of the most grounded forms of self-awareness. It's also one of the most stabilizing. Because when energy moves cleanly, thought becomes clear. And when thought becomes clear, everything else follows.

Coherence and the Body-Mind Field

The body and the mind are not separate systems. They're two aspects of the same process. What happens in thought affects the body. What happens in the body affects thought. This isn't philosophy — è fisiologia. Stress changes your heartbeat. Breath changes your mood. A single memory can tighten your chest. A moment of calm can slow your thoughts.

This connection is not just mechanical. It's energetic. It forms what può essere chiamato a body-mind field — an integrated system where electrical, chemical, and perceptual signals constantly interact. This field is always active. You don't have to think about it. But once you start noticing how it behaves, you can begin to influence it — not by force, but by rhythm.

At the center of this field is a principle called coherence. Coherence is a state in which the different systems of the body — brain, heart, breath, nervous system — fall into a synchronized rhythm. They're not just functioning. They're aligned. When coherence increases, energy moves more efficiently. Perception becomes clearer. Thought slows down, but gains depth. The body feels grounded. The mind feels quiet.

This state isn't rare. It can be induced through simple actions — slow breathing, deep focus, appreciation, or even by standing still with attention placed on the heart. The heart, in fact, plays a central role. It generates a measurable electromagnetic field, one that shifts according to emotional state. When you're stressed, the pattern becomes erratic. When you're calm, it becomes smooth and rhythmic. These shifts affect the brain, which responds in turn.

Studies in biofeedback and neurocardiology have shown that coherence improves cognitive function, decision-making, and emotional regulation. When you're coherent, the brain processes information more clearly. You're less reactive. You feel connected to your body. You think better — not faster, but cleaner.

Coherence is not relaxation. It's not passivity. It's alert stillness. You're present, but not tense. Focused, but not rigid. This state supports performance under pressure, recovery from stress, and clearer interpersonal interaction. It also opens a kind of perception that isn't just intellectual — it's intuitive.

When the system is incoherent — when parts of the body are out of sync — attention scatters. Thought fragments. You may feel wired but unfocused, tired but restless. This doesn't always come from big emotions. It can result from constant low-level stress, poor sleep, scattered attention, or unprocessed emotion. Over time, incoherence becomes a background state — so familiar it feels normal.

But the body remembers what coherence feels like. And with practice, it returns. The key is not force. It's rhythm. Regular breathing. Grounded posture. Attention moved gently inward. Each of these creates conditions where systems begin to align again. When the mind listens to the body, instead of overriding it, the field begins to stabilize.

Coherence also deepens perception. In this state, you're more likely to notice subtle shifts — in mood, in space, in the energy of a conversation. You don't have to interpret everything. You just register it more fully. This kind of awareness isn't hypervigilance. It's presence without strain. The system is alert, but not in defense.

This state also supports clearer boundaries. When you're coherent, you're less likely to absorb others' tension, confusion, or emotional projection. You can sense it without taking it on. Your system has more stability, more space, and more choice.

Coherence doesn't mean perfection. You'll still feel stress, frustration, fatigue. But you recover faster. You don't get pulled as deeply into loops. And when you do, you notice it sooner. That awareness becomes the first step in returning to balance.

One of the simplest ways to increase coherence is through heart-focused breathing — slow, steady breath directed toward the chest area, with attention placed not on thought, but on the sensation. Adding a feeling of calm, care, or appreciation strengthens the effect. It sounds simple, and it is. But repeated regularly, it changes the system — not just emotionally, but neurologically and energetically.

The body-mind field is always in motion. But motion can be harmonious or chaotic. Coherence is not about freezing that motion. It's about bringing it into rhythm. And when the field becomes coherent, experience becomes clearer. What was foggy becomes sharp. What was reactive becomes responsive. And from that state, you can act — not from pressure, but from precision.

Frequencies of Emotion and Environmental Influence

Emotions don't just affect the inner world. They radiate outward, shaping the space around us in ways that are often subtle but unmistakable. When someone walks into a room angry, the shift can be felt before a word is spoken. When someone is deeply calm, others tend to settle. This influence isn't about dramatic behavior. It's about frequency — the vibrational tone each emotional state carries.

These frequencies are not metaphorical. They manifest through real physiological changes: breath rate, muscle tension, heart rhythm, facial expression, and posture. The body becomes a transmitter, constantly broadcasting signals that others unconsciously register. These signals influence mood, perception, and even thought – not only in individuals but in groups.

You can feel this in a crowded train where everyone is anxious. You can feel it in a quiet room where someone has just cried. You can feel it in a conversation that hasn't turned hostile, but carries tension just beneath the surface. These aren't just social cues. They're energetic fields shaped by emotion, felt directly by the nervous system.

When emotional frequency is clear and stable, the environment tends to reflect that. Focus deepens. Interaction becomes smoother. Thought clarifies. But when emotional energy becomes chaotic – fragmented by stress, frustration, or reactivity – the field becomes noisy. Communication scrambles. Attention scatters. People pick up on it even if they can't explain what's wrong.

This isn't mystical thinking. It's direct experience. Emotional states, especially when repeated over time, imprint themselves into spaces. Just as a physical room accumulates sound and scent, a social space accumulates tone. That tone becomes part of the environment, shaping how others feel and behave within it.

In group settings – work, family, education – emotional frequency spreads quickly. Not through ideas, but through entrainment. One person's coherence can help stabilize others. One person's unchecked agitation can shift the entire group into tension. These patterns aren't always visible, but their effects are measurable. They change heart rhythms, cortisol levels, and behavioral responses in those nearby.

What's often overlooked is the role of awareness in this process. The more conscious you are of your own state, the more choice you have in shaping the field around you. When you feel reactive, and you catch it early, you can shift. That shift isn't just for your own benefit. It affects others. Sometimes dramatically.

Even physical environments hold emotional charge. A space where difficult conversations happen regularly tends to feel heavy. A space where attention, care, and presence are consistently practiced feels more open. These impressions are not imaginary. The body picks up on subtle cues in lighting, sound, layout — but also in energy: the unspoken atmosphere created by what has been felt there.

Caring for your environment, then, goes beyond cleanliness or aesthetics. It involves energy hygiene — the intentional management of what emotional frequency you bring into a space, and how you leave it behind. This can be as simple as pausing before entering a room. Resetting your breath. Letting go of tension. Or checking the emotional tone before speaking.

The same applies to communication. Every word carries frequency. Not just based on meaning, but on how it's said — the pace, the charge behind it, the intent. People respond less to content than to tone. A well-phrased sentence filled with irritation is received as aggression. A quiet word from a calm mind carries more weight than a flood of forced positivity.

This isn't about suppressing emotion. It's about taking responsibility for your signal. You don't have to be perfect. You just have to be aware. Emotional frequency changes throughout the day. But certain states become habitual. Some people live in low-level anxiety without noticing. Others operate from subtle defensiveness, irritation, or withdrawal. These states are contagious — not because they're loud, but because they're persistent.

Shifting frequency doesn't mean denying emotion. It means being honest with it — feeling it clearly, without feeding it. Emotions move when allowed to. They stagnate when resisted or exaggerated. When you name the state and stay with it — without turning it into a story

— it begins to shift. Energy that was locked starts to move. And the field clears.

This is how emotional resonance works. You don't just regulate yourself. You attune to the moment. You recognize what tone is needed, what would stabilize the space, what would bring clarity. And you adjust — not to please others, but to stay aligned with what feels real and balanced.

The space you move through — whether public or private — is influenced by your emotional field. And in turn, that space influences you. Learning to track and adjust that exchange is one of the quietest, most practical forms of energy work. It doesn't require belief. Just attention.

Emotional Resonance and Energy Hygiene

Some people walk into a room and shift the atmosphere without saying a word. Others speak for hours without changing a thing. The difference often has nothing to do with what's said, but with emotional resonance — the way someone's internal state affects others on a nonverbal, energetic level.

Resonance happens when systems fall into rhythm with each other. In music, it's when strings vibrate in harmony. In physics, it's when frequencies align and amplify. In human interaction, resonance occurs when emotional states synchronize — intentionally or not. When someone is centered, clear, and grounded, their presence brings others into a more stable state. When someone is chaotic or emotionally charged, others tend to absorb or reflect that frequency.

We're always broadcasting something. It's not about performance, charisma, or control — it's about coherence. If you're angry but pretending to be calm, people don't trust the signal. If you're present, even in silence, others feel it. The nervous system responds before the mind interprets.

Emotional resonance isn't just about impact — it's also about sensitivity. The more attuned you are to subtle emotional shifts, the more likely you are to pick up what's happening around you. This is a strength, but it also requires boundaries. Without clarity, emotional sensitivity can lead to overwhelm, absorption, or confusion.

This brings up a key point: resonance is two-way. You're not just influencing — you're being influenced. Your emotional field is porous. It takes in signals constantly, especially in close contact or crowded environments. Over time, this can create emotional residue — low-grade tension, fatigue, or agitation that doesn't seem to come from inside but feels undeniably present.

That's where energy hygiene becomes essential. Just like physical hygiene keeps the body clean and functional, energy hygiene keeps the emotional system clear and stable. It's not about blocking out the world or staying neutral. It's about staying clear enough to notice what's yours and what's not — and knowing how to reset when the system gets cluttered.

Basic forms of energy hygiene are simple but powerful:

- **Pausing after social contact** to reset your system, especially after intense conversations.
- **Conscious breathwork** to discharge tension and re-center.
- **Physical movement** — walking, shaking out the body, even stretching — to help release stagnant emotional energy.
- **Brief silent presence**, not as escape but as recalibration — a return to your own signal.

Also important is **attention management**. What you focus on feeds your emotional field. If your attention is pulled constantly by news, arguments, or emotional noise, the system becomes saturated. Clarity drops. Your baseline emotional tone shifts, often without you realizing it. Protecting your focus is not selfish — it's functional. It keeps your field strong, so that your presence has integrity.

Another aspect of emotional hygiene is discernment. Not every emotional wave you pick up requires engagement. Some aren't yours. Some aren't relevant. Sensitivity doesn't mean absorption. It means recognizing what's happening without getting pulled into every frequency that passes through your awareness.

Boundaries support this. Not walls — but clarity. When you're clear about your state, your needs, and your limits, you're less likely to collapse into other people's emotional fields. You can stay open without being porous. This isn't about control — it's about stability.

There's also a practice of emotional reset. At the end of the day, it can be useful to ask: what emotional energy am I carrying that isn't mine? What do I want to release? Even a brief check-in like this helps prevent accumulation. It creates space — not just mentally, but energetically.

Over time, developing strong emotional hygiene increases resilience. You become less reactive, more perceptive, and more available. You're not as easily pulled by collective moods or interpersonal tension. Your emotional field becomes something you can trust — not because you're always happy, but because you're consistently aware.

This doesn't make you immune to stress or conflict. It just gives you more room to respond instead of react. It allows you to hold space for others without losing your center. And it helps make your presence a stabilizing influence — not just for those around you, but for yourself.

In a world that constantly pulls attention outward, emotional resonance and energy hygiene become quiet forms of mastery. They don't require intensity or control. They require attention, honesty, and practice. And through them, the emotional body becomes not just a reaction system, but a source of clarity, balance, and intelligent connection.

Chapter 8: Intention, Attention, and Reality Perception

What you focus on becomes more real. Not just in terms of thought, but in terms of perception, emotion, and action. The mind is constantly filtering reality — not showing you everything, but only what matches the patterns it's already tracking. Attention is the filter. Intention is the direction.

Together, they create what you experience as real. Not in the philosophical sense of inventing the world, but in the practical sense of shaping how the world appears, what options you see, what information you notice, and how you respond. This isn't imagination. It's selection.

This chapter explores how intention and attention work — not as vague concepts, but as precise forces that shape the field of awareness. Understanding their structure brings clarity. Using them consciously brings power.

The Mechanics of Attention

Attention isn't thought. It's the spotlight in which thought appears. It moves faster than logic, faster than memory. It shifts constantly — pulled by noise, desire, interruption, or urgency. And in that shifting, it decides what becomes visible and what fades into the background.

The mind can only process a limited amount of information at a time. Out of billions of stimuli, only a few reach conscious awareness. What makes the cut is determined largely by where attention is placed. That means attention functions as a gatekeeper — deciding what enters, what stays, and what is ignored.

Attention is not neutral. It's shaped by habit, emotion, and belief. If your system is trained to look for threat, it finds it – even where it's not present. If your focus is anchored in fear, doubt, or comparison, the world will seem like a mirror of those states. But if attention becomes stable, direct, and calm, perception shifts. Possibility becomes more visible.

There's a common misunderstanding that attention is something you control directly. But in most people, attention is reactive. It follows the loudest signal – an incoming message, a sharp sound, a sudden thought. This reactivity fragments perception. The mind jumps from one thing to another without settling.

Stabilizing attention doesn't mean holding it by force. It means training the system to reduce unnecessary noise so that attention can land where it's most useful. This takes practice. Breath work, meditation, deep listening, or any form of intentional stillness strengthens the system's ability to stay with one object without distraction.

One of the overlooked facts about attention is that it feeds whatever it touches. What you place your attention on becomes more active in the system. If you focus on pain, pain intensifies. If you focus on a solution, new paths begin to emerge. Attention adds weight. It gives shape.

This has consequences in daily life. If your attention is constantly pulled by conflict, you're likely to feel surrounded by it. If it's pulled by anticipation, you may live in perpetual preoccupation. The outer world hasn't changed – but the part of it you're tuned to has.

Attention can also be selective in dangerous ways. When it becomes locked – through fixation, obsession, or unresolved emotion – it stops updating. You keep seeing the same pattern even when reality has moved on. The brain creates a closed loop. Fresh information doesn't enter. Attention reinforces the loop instead of breaking it.

Training attention means widening its flexibility. You learn to focus when needed, release when it's time, and shift deliberately when the moment changes. This kind of flexibility improves mental clarity, emotional balance, and creative response.

There's also a difference between narrow and open attention. Narrow attention is useful for tasks, detail, precision. But it creates tension if held too long. Open attention — a panoramic, receptive state — allows deeper perception, better intuition, and more space for insight to arrive. Moving between these two modes, without getting stuck in either, is part of healthy cognition.

Attention also influences physiology. When you focus on the breath, the breath changes. When you place attention in the body, tension often releases. This isn't imagination. It's interaction. Attention is a form of contact — and contact brings change.

Finally, attention is relational. What you give attention to in others grows. When you focus on their potential, their presence, or their coherence, they often respond in kind. When your attention reinforces only their fear or confusion, they contract. This doesn't mean you're responsible for others — but it shows how attention carries impact, silently and constantly.

Understanding the mechanics of attention is a step toward reclaiming choice. You don't have to chase every thought. You don't have to react to every signal. With awareness and practice, attention becomes less of a reflex and more of a tool. A way to bring clarity to noise. A way to return to what matters.

Intention as a Directional Force

Intention is not wishful thinking. It's not about repeating mantras or holding abstract desires. It's a directional force — a subtle but powerful alignment between thought, energy, and action. When intention is clear, it organizes attention. It shapes perception. It sets a tone that affects everything that follows.

Unlike desire, which often comes from a place of lack, intention arises from clarity. It doesn't chase. It directs. It acts not out of pressure, but from alignment. This is what gives it strength. It doesn't depend on the external environment. It moves from within, but it moves cleanly — without struggle, without grasping.

In practical terms, intention acts like a stabilizing axis. When it's strong, attention returns to it automatically, even when distractions pull. You're less likely to get lost in noise because something deeper is holding the system together. This internal direction creates consistency — not rigid control, but a kind of inner gravity.

Many people confuse intention with mental effort. But the two are different. Effort pushes. Intention guides. You can act with enormous intensity and still be unclear about your intention. You can also move with quiet precision, held together by a clear inner direction. The difference is not in volume, but in alignment.

This kind of alignment shows up most clearly in high-pressure situations. When someone moves with clear intention, others feel it. Their actions have weight. Not because they're loud or forceful, but because there's no doubt in the system. The internal message and the external action match.

You can sense it in communication too. Words spoken from clear intention land differently. They don't try to persuade. They reveal something stable underneath the surface. Even silence, when held with intention, can be more effective than explanation. It carries signal.

But intention can also be misused. When it's driven by fear, insecurity, or control, it becomes manipulative. It may still be effective — but it distorts the field. You might get what you want, but you create residue: tension, resistance, or confusion in others. The real measure of clean intention is that it creates clarity, not just outcomes.

That's why understanding your own intention — beneath the surface — is critical. It requires honesty. Not just "What do I want?" but "Why do I want it? Where is this moving from?" Often, the most important shift isn't changing the intention, but cleaning it up. Removing the noise. Releasing the grasping. Returning to something more steady.

There's a deeper layer, too. In many traditions, intention is seen not just as a mental decision, but as a tuning. A way of aligning the self with a larger pattern – whether that's called nature, life, the field, or something else. In this sense, intention doesn't override what's happening. It synchronizes with it. You're not forcing events. You're moving in rhythm with them.

In this model, intention becomes less about control and more about resonance. You set a tone – not just with your thoughts, but with your posture, breath, and emotional field. That tone informs your actions. It shapes what you notice. It affects how others respond. The external world doesn't obey you – but it begins to reflect the signal you're holding.

This is not magic. It's feedback. The mind, the body, and the environment are in constant exchange. When your signal is clean and your direction is stable, the feedback becomes clearer. You begin to notice the difference between distraction and relevance, between reaction and response.

It's also worth noting that intention doesn't always lead to what you imagined. Sometimes it clears the path by removing what no longer fits. Sometimes it leads to discomfort before alignment arrives. But even in those moments, a strong intention holds you steady. It keeps you from collapsing into confusion when conditions shift.

Setting intention can be simple. It can be as direct as a sentence: "Move from clarity." "Speak from calm." "Stay with truth." The words don't matter as much as the state they represent. And once the signal is set, you return to it. Again and again, until the system begins to follow it naturally.

When intention becomes steady, it doesn't need reinforcement. It shows up in action, tone, timing. Others feel it. You feel it. You're not forcing your way through life. You're aimed – quietly, precisely – and you let that direction do the work.

Perception Filters and the Construction of Reality

You don't see reality as it is. You see it filtered — shaped by memory, emotion, belief, and expectation. The brain is not a passive recorder. It's an active editor. It selects, organizes, interprets. Every moment, it throws out most of what it receives and keeps only what matches its internal model of what matters.

This model is built over time: through culture, trauma, experience, repetition. It becomes your lens. And once the lens is in place, it keeps reinforcing itself. You see what you already believe. You overlook what doesn't fit. This is not a flaw — it's how the system protects efficiency. But it creates a problem: reality becomes a closed loop.

Two people in the same situation can have completely different experiences — not because the world changed, but because their filters did. One sees opportunity; the other sees threat. One hears feedback; the other hears attack. Same words, different world.

These filters operate mostly beneath awareness. You're not choosing what to believe in the moment — you're reacting based on past inputs. This means that perception is often automated. You think you're seeing what's there, but you're actually seeing what you're trained to see.

One of the strongest filters is identity. If you believe you're weak, your mind will highlight failure, threat, and rejection. If you see yourself as competent, the same events might register as challenge or feedback. The external facts don't change — the interpretation does. Identity colors perception.

Another major filter is emotion. In a state of fear, your attention narrows. You look for risk. You misread neutral signals as hostile. You miss nuance. When calm, the same situation appears more open, more manageable. Emotion doesn't just affect mood — it reshapes the architecture of what feels possible.

Language is a filter too. The words you use internally define the categories you see. If your inner vocabulary is filled with "should," "never," "not good enough," or "I always," you trap experience into fixed patterns. You stop perceiving things as they are, and start perceiving them as how they fit your script.

Attention reinforces these filters. Where you look becomes what you see. The more often you notice a certain pattern — unfairness, chaos, disrespect — the more dominant it becomes in your internal map. You begin expecting it. You start missing everything that contradicts it. And in time, it stops being a perception. It becomes your version of reality.

But filters can change. Not by force, but through awareness. The first step is noticing what you assume. Catching the moments when your reaction doesn't match the facts. Asking: what am I actually seeing here — and what am I projecting?

This doesn't mean denying your experience. It means updating it. Recognizing that perception is fluid. That the map is not the territory. That every story you carry about yourself and others is a compression of complexity. And that it can be restructured — gently, with attention.

One useful practice is zooming out. When you feel locked into a perspective — anger, fear, certainty — pause and widen the frame. What's another way this could be interpreted? What am I not seeing? Not to erase your point of view, but to add space around it. This space is where freedom begins.

Another practice is contradicting your own filter. If you're always expecting rejection, look for signs of welcome. If you assume people won't listen, test what happens when you speak clearly. You don't need to fake belief. You need to create friction in the pattern. Enough to break the automatic loop.

Also important is who you spend time with. The emotional and perceptual filters of those around you shape your own. If you're always in environments of complaint, panic, or cynicism, your lens adjusts accordingly. Not by agreement, but by exposure. Filters sync through contact.

The goal is not perfect perception. That's not possible. The goal is flexible perception. A system that updates, adapts, and stays close to what's actually happening — not just what's been conditioned to appear. This requires discomfort. Letting go of old maps. Feeling uncertainty. Moving without a fixed script.

But the payoff is clarity. When filters loosen, reality becomes less threatening. More dynamic. More nuanced. You don't need to control everything. You just need to see what's actually there — and move from that.

Perception is not passive. It's creative. And when you begin to take responsibility for how you see, your reality starts to respond in kind.

Directed Focus and Conscious Reality Shaping

Focus is not just about productivity. It's about reality selection. Where you place your attention — consistently, deliberately, and with clarity — becomes the version of reality that grows. Not because the world bends to your will, but because your nervous system, your perception, and your actions all start to align with the field you're reinforcing.

Directed focus is the act of choosing what you give your energy to. This doesn't mean ignoring problems, denying pain, or pretending. It means deciding what matters enough to keep alive inside your awareness. Everything you focus on is being fed — emotionally, neurologically, and energetically.

This is happening all the time, whether you're aware of it or not. Most people move through life reacting — to noise, pressure, urgency. Their focus is pulled, not placed. It flickers between inputs, never stabilizing long enough to form depth. In that state, perception becomes fragmented. Power is diluted. Presence fades.

Directed focus changes that. It creates a point of orientation. It doesn't mean staring at a goal obsessively. It means maintaining a steady alignment with a chosen state, value, or direction. For example: clarity instead of confusion. Depth instead of speed. Listening instead of defensiveness. These aren't just ideas — they're positions you return to, again and again, as the world pulls you off balance.

When focus is consistent, patterns start to shift. The brain forms new pathways. The emotional body adjusts its tone. The environment responds — not always dramatically, but subtly. You start noticing different opportunities. You speak differently. Others relate to you differently. What seemed fixed begins to open.

This doesn't happen through force. Forced focus creates tension. It becomes a struggle against distraction. Directed focus is different. It works like gravity: a steady pull, not a rigid grip. It holds the system in alignment without shutting it down. It has flexibility. If something urgent arises, you respond. Then you return.

There's also a difference between focusing on outcome and focusing on quality. Most people chase results — numbers, validation, completion. But this kind of focus is often restless. It keeps the system in a state of "not yet." In contrast, focusing on the *quality* of being — how you move, how you relate, how you hold attention — shifts everything into the present. The moment becomes the field of change, not just a step toward something else.

This kind of focus doesn't ignore the future. It includes it — but it brings power back to now. And in doing so, it changes how future moments unfold. You're not just preparing for impact. You're already shaping it through your current signal.

A big part of conscious reality shaping is learning to *withdraw attention from what drains you* — not as avoidance, but as discipline. Every time you engage obsessively with fear, conflict, or resentment, you reinforce it. Not because it's wrong to feel those things, but because prolonged attention hardens them into identity. When you remove focus from what no longer serves, it begins to lose momentum.

This is not repression. It's selection. You feel fully, but you don't stay trapped. You observe, but don't feed. You choose what to keep active. And that choice — repeated consistently — starts to shift what shows up in your space.

This is where focus becomes more than internal. It becomes environmental. What you focus on changes how people experience you. It alters how they respond. It influences which conversations emerge and which dissolve. Directed focus becomes a quiet organizing force — not loud, not forceful, but undeniable.

One of the simplest ways to test this is in conversation. Choose to stay focused on clarity — not on being right. Or focus on presence — not performance. Notice how the tone changes. Notice how your nervous system stabilizes. The more practiced your focus, the less reactive your environment tends to become.

This doesn't mean you control the world. It means you shape the field you interact with. You create conditions — internally and relationally — that support different kinds of experience. And over time, those conditions become your baseline.

Directed focus also simplifies complexity. When the system is overwhelmed, the mind spirals. But when you return to a chosen point — breath, sensation, question, value — noise drops. The signal strengthens. And from that still point, movement becomes more precise.

This is what conscious reality shaping means. Not escaping the world, not rewriting it with fantasy, but selecting the version of it you engage with — deliberately, repeatedly, and with awareness. The world you live in begins with what you focus on. Not as a metaphor, but as a function of perception, emotion, and action.

Your attention is not passive. Your focus is not neutral. They build the lens through which experience is formed. And when you start using that lens consciously, you stop living at the mercy of what appears — and start participating in what unfolds.

Chapter 9: Altered States and Non-Ordinary Perception

Consciousness is not a fixed setting. It shifts — quietly, constantly — in response to rhythm, environment, emotion, intention, chemistry, and focus. Some states are ordinary: waking, sleeping, thinking, reacting. Others are altered: dreamlike, expanded, hyper-attuned, still beyond thought.

These altered states are not escape routes. When grounded and used with clarity, they can open access to new forms of perception — of time, identity, memory, and meaning. Some arrive through practice. Others through intensity. Some are triggered by silence, breath, movement, or even trauma.

This chapter explores the terrain of non-ordinary states — how they work, what opens them, and how to relate to them without getting lost or dependent. Not from a mystical angle, but from direct observation of how consciousness behaves when the usual filters begin to shift.

Shifting States of Consciousness

You're never in one fixed state for long. The mind flows through stages across the day: alert, scattered, calm, dull, focused, reactive. Most of this goes unnoticed — until something changes sharply. A deep breath. A shock. A moment of silence. Suddenly, the world feels different. Thought slows. Time stretches. You see from a slightly new angle.

These shifts are not exotic. They're built into the nervous system. You experience one version of reality when rushing, another when still. One version when angry, another when in deep listening. The outer world stays the same. But your state alters what you're able to perceive — what details you notice, what connections you draw, what you feel in your body.

This flexibility is natural. But in most people, it's unconscious. The state changes, but there's no awareness of the shift. The more consciously you track your own state, the more access you have to choice. You don't control what arises, but you can learn to influence how long you stay in a state, and how you move out of it.

Some altered states arise spontaneously — through music, nature, solitude, exhaustion, or crisis. Others can be cultivated intentionally through breathwork, sensory deprivation, fasting, stillness, or focused attention. These aren't necessarily mystical techniques. They simply reduce external noise and internal chatter, allowing the system to stabilize and reorganize.

In many traditions, these states are not seen as abnormal, but as part of the spectrum of human experience. They're approached with respect, not worship. They reveal possibilities of mind that are always present, but usually drowned out by the pace of ordinary life.

One of the most immediate ways to shift state is through the body. Slowing the breath, relaxing the jaw, grounding the feet — these cues signal safety to the nervous system. When the body feels safe, consciousness opens. It's not about entering a trance. It's about exiting contraction.

Another gateway is attention without content. When you sit with awareness placed on the space between thoughts — not forcing, not drifting — the usual mental activity starts to fade. What remains is presence. Not passive, not sleepy. Just alert, and still. In that stillness, new forms of perception sometimes appear: spatial awareness, intuitive sensing, emotional clarity without analysis.

Time behaves differently in these states. It can dilate, collapse, or lose structure entirely. You may feel minutes stretch like hours, or hours pass in what feels like a few moments. This isn't illusion. It's a change in how the brain organizes experience when usual reference points are removed.

Memory also shifts. In altered states, past experiences can surface in nonlinear ways. Sometimes without emotional charge. Sometimes with new insight. This isn't about reliving the past. It's about accessing it from a different angle, free from the framing that kept it fixed.

These changes can be beautiful, but also disorienting — especially if they happen without context. That's why integration matters. A powerful experience means little if it doesn't inform your life. Insight that doesn't translate into behavior fades. The goal isn't just to access altered states — it's to bring back something usable.

And not all altered states are expansive. Some are contracted: dissociation, numbness, hyper-alert survival modes. These are also altered states — but ones born from stress, not stillness. The nervous system narrows perception to preserve energy or avoid overwhelm. These states can mimic calm, but they lack connection. The mind may be quiet, but it's shut down, not open.

Discerning the difference is part of the work. Real expansion feels grounded. Even when intense, it contains clarity. There's a sense of returning, not escaping. A quiet inner "yes" that confirms coherence rather than confusion.

Ultimately, altered states aren't the goal. They're tools — not in the sense of performance or technique, but as natural capacities of the human system. When approached with respect and awareness, they offer perspective. Not as escape, but as contrast. A reminder that the version of self and world you inhabit moment to moment is flexible, not fixed.

Learning to move between states — and to stabilize in clarity rather than chase stimulation — opens new ground. Not for spectacle, but for deeper attention, cleaner action, and more honest perception.

Access Points: Breath, Sound, and Stillness

Altered states don't always require intense practice or dramatic intervention. Some of the most effective shifts in consciousness happen through simple, accessible gateways — breath, sound, and stillness. These aren't techniques to master. They're natural systems already operating in the body. The key is how you use them.

Breath is the most immediate access point. It sits at the intersection of the conscious and unconscious. You breathe automatically, yet you can also choose how you breathe. That makes it a rare kind of bridge — one that gives you direct influence over the nervous system.

The pattern of your breath reflects your state. When you're anxious, it's shallow, fast, and high in the chest. When you're calm, it's deeper, slower, grounded in the belly. Most people never observe this connection. They live with the breath stuck in a reactive loop. But once you begin to notice and adjust it, the entire internal field starts to shift.

Slowing the breath — even slightly — can reduce heart rate, soften mental noise, and open perception. You're not forcing relaxation. You're creating conditions that allow the body to remember how to rest. With enough repetition, the system starts to anticipate the shift. Breath becomes signal, not just function.

Even small practices — three deep breaths before entering a room, steady nasal breathing during stress, exhalation extended slightly longer than inhalation — can reset the nervous system. You're not escaping life. You're re-entering it with more presence.

Sound is another entry point. The body is built for resonance. Every cell vibrates. Every organ has a frequency range. Sound interacts with the nervous system directly, bypassing language. It can settle, stimulate, disrupt, or entrain — depending on what frequencies are used and how they're received.

Music is the most obvious form, but not the only one. Repetitive sounds — drumming, humming, chanting — can create rhythmic stability in the brain. They engage the auditory system in ways that reduce thought loops and help shift the brain into different wave states (alpha, theta, delta), where access to non-linear perception increases.

Even single tones — like a singing bowl, a low hum, or a sustained vowel — can create immediate change. When done with attention, sound becomes a container. It holds focus. It slows distraction. The vibration moves through the body and creates a subtle, physical coherence that isn't forced — it's felt.

Silence, too, has sound. True silence is rare. But when the external volume drops low enough, inner sounds become audible — breath, heartbeat, ambient noise. This shift in attention recalibrates perception. The mind, starved of external input, begins to widen. Space opens. Thought slows.

That leads to the third access point: stillness. Stillness is not the absence of motion. It's the absence of unnecessary motion. It's what remains when you stop reacting. You sit, stand, or lie down — not as an idea, but as a direct experience. The body doesn't move. The breath quiets. The eyes settle. Attention begins to collect.

In stillness, the usual distractions have nowhere to hide. You begin to notice how much of your restlessness is habitual. How much comes from stimulation, memory, anticipation. When these settle, even briefly, a different kind of awareness surfaces — one that doesn't reach, doesn't comment, doesn't struggle. It just registers.

Many traditions use stillness as a core method — not because it's passive, but because it sharpens perception. It brings coherence to scattered systems. It reveals what's usually hidden beneath action: fatigue, confusion, clarity, grief, insight. You don't have to chase anything. You just stop interfering.

Combining these access points — breath, sound, and stillness — amplifies their effect. A quiet room, steady breath, and a low hum can shift consciousness in minutes. Not because you're escaping the mind, but because you're giving it a new rhythm to follow. The system wants coherence. It just needs an invitation.

What's important is not turning these access points into rituals or performances. It's about relationship. Your breath is always with you. Your sense of tone and rhythm is always active. Your capacity to be still is always available — even if only for ten seconds. These are not spiritual tools. They're human ones.

Using them doesn't mean rejecting ordinary consciousness. It means expanding your range. Giving yourself more options. So that when stress rises, you don't collapse. When noise floods in, you don't drown. You shift. You tune. You respond from a different place.

Over time, this changes your baseline. You stop living in the outer edge of your nervous system — constantly stimulated, reactive, fragmented. You move closer to center. Not all the time. Not perfectly. But reliably enough to build trust in your own capacity to reset.

Access points don't solve everything. But they remind the system that another way of being is always near — not at the edge of some long path, but right here, just below the surface of noise.

Dangers of Escapism and the Discipline of Integration

Non-ordinary states can be beautiful, intense, even life-altering. They can dissolve boundaries, bring deep insight, and shift the emotional system in ways that regular thought cannot reach. But they also come with a risk — the temptation to escape. To chase the state instead of letting it serve the deeper movement of awareness.

This is where a quiet discipline becomes essential: integration.

The human mind loves peak experiences. They break the monotony. They promise transformation. They feel like proof that something deeper is real. But a peak without integration is just another high. It may feel profound in the moment, but if it doesn't translate into how you speak, act, relate, and choose — it fades. Or worse, it creates a split between "ordinary life" and "real self."

Escapism shows up in subtle ways. Obsessing over altered states. Constantly seeking silence or "energy." Avoiding conflict in the name of being above it. Refusing to engage with responsibility or relationships because they feel too "dense." These patterns are not evolution. They're defense — disguised as spirituality, or growth.

A true altered state — when real and coherent — doesn't pull you away from life. It sharpens your contact with it. You become more honest. More precise. More available. Less reactive. You don't need to declare that you've changed. People can feel it. And the change shows up in how you deal with difficulty, not just in how you feel when things are still.

That's the discipline of integration: translating insight into structure.

It means asking real questions:– How does what I experienced change how I relate to others?– What does it shift in how I respond under pressure?– What behavior no longer fits, even if it's familiar?– What does this insight require me to do next?

Integration doesn't always feel exciting. Sometimes it's quiet. Sometimes it's hard. Sometimes it means making very ordinary changes: how you use your time, how you handle money, how you listen. But these small shifts are the container that holds the signal of transformation. Without them, the insight leaks. With them, it takes root.

The nervous system, too, needs time to catch up. Deep states — especially if intense — can overwhelm the body if not grounded. Rest is part of integration. So is movement, hydration, conversation, and basic contact with reality. These things don't dilute the experience. They give it shape.

Another part of integration is the ability to let go of the experience itself. Not every insight needs to be turned into a story. Not every vision has to become a project. The need to hold onto peak moments often reveals fear: fear that clarity will fade, that depth won't return, that without a special state, you're just lost again.

But insight that is real doesn't need constant reinforcement. It changes your trajectory without clinging to the event that started it. You don't have to hold the memory. The system remembers the shift.

Integration also means confronting contradiction. The parts of you that resist the change. The relationships that no longer reflect your new alignment. The habits that pull you back into old loops. These moments are not failures — they're invitations to make the internal shift external. Not with drama. With action.

This is why community matters. Not one built on shared belief, but on shared practice. People who care less about how "expanded" you felt, and more about how you live now. Who reflect back your behavior, not your story. Who can meet you in the ordinary with the same presence as in the peak.

In the end, altered states are part of a much larger process — not the goal, but a window. What matters is what happens afterward. Not what you saw, but how you move now. Not what you understood, but what you choose when clarity fades.

Without integration, altered states become another form of escape. With integration, they become fuel — quiet, deep, and steady — for a life that's not just more conscious, but more real.

Grounding and Returning from Expanded States

Coming back from an altered state can be disorienting. The expanded field of perception collapses. The clarity fades. Normal life reappears — full of noise, tasks, unresolved tension. The mind tries to hold on to what just happened, but the body is already shifting back into old rhythms.

This return is one of the most delicate and often neglected phases of inner work. The experience was clear, maybe even powerful. But now you're left with a question: what do I do with this in the middle of a conversation, in traffic, at work, in the ordinary?

The transition between states matters. How you come back determines whether the insight integrates or gets buried. Rushing back into distraction, noise, or emotional charge can cause confusion. It can even create a backlash — where the nervous system, overwhelmed by the contrast, rejects the entire experience as unrealistic.

That's why grounding is essential.

Grounding doesn't mean returning to old patterns. It means reconnecting the expanded state to something real, tangible, and usable. It anchors the shift in the body. It helps stabilize the nervous system, so the insights don't just float — they settle. They begin to live in your posture, your tone, your timing.

Grounding can be simple. Touching something solid. Drinking water slowly. Walking barefoot. Taking three conscious breaths. Speaking one clear sentence. The point isn't ritual. It's contact — with the moment, with your environment, with your body.

It also means re-engaging deliberately. Instead of snapping back into ordinary rhythm, you re-enter with care. You listen to what your body is doing. You notice your language. You allow space for your inner field to adapt before jumping into stimulus. This keeps the nervous system from fracturing the shift into "before" and "after."

Another aspect of grounding is letting go of the need to interpret immediately. You don't have to explain the experience to yourself or others. You don't need to decide what it meant right away. Often, the mind rushes to define something so it can feel in control again. But some insights need silence. Some shifts only reveal their meaning slowly, through action, not reflection.

During the return, it's also normal to feel emotionally raw. Expanded states often open emotional material that hasn't been fully processed. Don't rush to close it down. But don't exaggerate it either. Let the system discharge what it needs to. Keep your breath steady. Avoid re-triggering old stories. Let the body move through it with minimal interference.

If the experience was intense — physically or emotionally — give yourself time. Rest. Hydrate. Eat something grounding. Avoid overstimulation. If you were alone, it can help to speak briefly with someone who listens well and doesn't try to analyze. Just being witnessed with presence can help bring the system back into balance.

If you were with others during the altered state, resist the urge to overprocess together. Sometimes too much talking immediately after can collapse the experience into narrative. Hold space for silence. Let meaning settle naturally over the next hours or days.

The main point is this: you don't have to hold onto the state. You don't have to preserve it, repeat it, or define it. What matters is what stays with you without effort. What continues to influence your perception, your choices, and your tone — even when the peak is over.

Coming back is not a loss. It's a return to form — with something new carried inside. You don't have to live in altered states to be awake. You have to let them inform how you live when you're not in them.

That's what grounding really is. Not anchoring to the past, but placing insight into structure. Letting expansion reshape how you move through daily life. Not by acting special. But by becoming more honest, more responsive, more present — quietly and without force.

When that happens, the space you touched in the altered state isn't gone. It's folded into how you see. It's in your timing, your eyes, your tone. You don't talk about it — you carry it. And that's what makes it real.

Chapter 10: Bodily Intelligence and Bioenergetic Fields

Most people treat the body as secondary – a vehicle for the mind, a container for experience, something to fix, decorate, or ignore. But the body is not passive. It doesn't just respond. It perceives, stores, remembers, and often knows before the mind does.

Every thought has a physical tone. Every emotion leaves a residue. Every choice is felt in the tissues. There is no real separation between mental, emotional, and physical states. They happen together, in a single field – expressed through sensation, posture, breath, and movement.

This chapter explores the idea of bodily intelligence – not as metaphor, but as a direct experience. It also introduces basic principles of bioenergetics: the study of how energy flows through the body, how tension is stored, and how awareness can shift not just the mind, but the entire system.

We're not building a theory. We're describing a living reality: that your body is already communicating – constantly. And the more you listen to it, the more access you have to clarity that doesn't depend on thought.

The Body as a Field of Intelligence

The body reacts before thought does. You walk into a room and feel discomfort without knowing why. You meet someone and instantly feel trust, or the opposite. These responses don't come from logic. They come from the body. From sensing.

This intelligence isn't mystical. It's sensory, neurological, energetic. The body tracks subtle signals: voice tone, micro-movements, spatial tension. It reads environment and emotion long before the conscious mind forms language around it. This is not intuition in the abstract. It's precision — based on deep, fast perception.

The problem is that most people live disconnected from this level. Trained to suppress sensation, to override feeling, to push through fatigue, to ignore tension until it becomes pain. The result is that the signals don't stop — they just get louder, or go underground.

The body stores what isn't processed. Not as memory in the mind, but as pattern in tissue: tight shoulders, shallow breath, clenched jaw, collapsed chest. These are not just habits of posture. They are maps of experience. They show what was endured, what was defended against, what was avoided.

Every chronic tension has a story — even if the story is forgotten. The tension remains. It shapes how you sit, how you speak, how you relate. It becomes part of your identity without your consent.

This is where body intelligence becomes important. The body doesn't lie. If you say you're fine but your stomach is tight, that tension holds more truth than your words. If you believe you're confident but your spine is curled inward, the body is revealing what's actually being lived.

Listening to the body doesn't mean decoding it like a puzzle. It means paying attention without interference. Not fixing. Not judging. Just noticing — with enough stillness to allow sensation to surface, and enough patience to let the system speak.

This kind of listening reveals that the body is not just reacting. It's trying to balance. Trying to discharge. Trying to return to coherence. When you give it space, it often knows what to do. Spontaneous sighs, small movements, tremors, shifts in breath — these are all forms of intelligence, not symptoms to eliminate.

The body also responds to intention. Not the kind of forced focus that tries to manipulate, but subtle direction. When you place awareness in a tense area and simply stay — without trying to relax it — the tissue often begins to soften. The breath changes. Energy moves. Something inside adjusts.

This interaction between awareness and the body is the foundation of bioenergetics. Not as a belief system, but as observation: energy flows where attention goes. Tension dissolves when pressure is removed. Presence restores movement where the system was frozen.

The body has its own timing. You can't rush release. You can't force openness. But you can stop interfering with it. And when you do, the system begins to self-regulate.

This is the essence of bodily intelligence. Not theory. Not exercise. Contact. With sensation. With signal. With the place inside you that knows what's happening — without needing to name it.

The more you build this contact, the less you rely on mental narratives to explain your state. You don't have to analyze everything. You can feel what's true. You can sense what's needed. And you can respond from a place that is not just reactive, but informed — by something deeper than thought.

Tension Patterns and Emotional Memory

The body remembers everything the mind forgets. Not in the form of detailed narratives, but through patterns: areas that stay tight, movements that never complete, postures that reflect old strategies of survival. These aren't just physical quirks. They are the residue of emotion — stored, frozen, and repeated.

Tension is often mistaken for strength. People carry it in their shoulders, jaw, abdomen, legs — unconsciously holding themselves together. But what's being held, exactly? Often, it's unexpressed emotion: fear that couldn't be felt, anger that had to be hidden, sadness that was never allowed space. The body took the charge and buried it in the muscle, the breath, the fascia.

These patterns are intelligent. They were the system's best solution at the time. A way to maintain function, protect vulnerability, avoid collapse. But over time, what was once adaptive becomes automatic. The body forgets how to let go — not because it can't, but because it doesn't feel safe enough to.

Each emotion has its own physiological signature.
- **Fear** tightens the gut, shortens the breath, narrows the eyes.
- **Anger** energizes the limbs, locks the jaw, raises the shoulders.
- **Grief** collapses the chest, softens the belly, weakens the knees.
- **Shame** pulls the spine forward, dims the face, silences the throat.

When an emotion is fully experienced, the body moves through it. The breath rises and falls, the energy peaks and resolves. But when expression is blocked — because it's unsafe, unwelcome, or overwhelming — the motion stops midway. What was meant to pass through becomes held. And held. And held.

Over years, these holding patterns become structural. The way you stand, the way you sit, even the way you speak. The muscles learn to contract preemptively, anticipating threat. The breath learns to stay shallow, avoiding sensation. The nervous system stays half-alert, never fully at rest.

This isn't dysfunction. It's memory.

And memory can be updated — but only through contact.

You can't talk a tension out of the body. You can't will it away through logic. The system needs to feel safe enough to soften, and that safety comes from presence. From attention that doesn't demand. From awareness that doesn't rush. From an environment where no part of you is being pushed away — not even the tightness itself.

Sometimes tension needs movement. Other times, stillness. Somctimes it needs to tremble, yawn, stretch, cry, breathe deeply. These aren't signs of weakness or regression. They're expressions of the body discharging what it couldn't finish earlier.

There's no fixed method. One person might find release through slow, mindful breath. Another through deep physical exertion. Another through touch. Another through vocal sound. What matters is not the form — it's the honesty of contact. Whether the movement is aligned with what the body is actually feeling, or imposed from the outside.

Working with tension patterns is not about "fixing posture." It's about reclaiming sensation. When you feel what's happening in your chest, belly, throat, spine — without judgment — you give the system permission to move. Movement doesn't always mean big gestures. Sometimes it's subtle: a change in breath, a drop in the jaw, a softening of the gaze.

Over time, as old tensions begin to release, something else becomes visible: underneath the pattern is the emotion that created it. Often in raw form — unfiltered, sometimes unfamiliar. It might be brief. It might be intense. But it's almost always clean. Not dramatic. Not loud. Just direct.

That's the point where memory becomes transformation. When the energy that was trapped is finally felt and allowed to move. You don't need to re-live every past experience. You don't need to know the story. You just need to let the system finish what it started.

The result is often subtle: more breath, more clarity, more ease in the body. You stand differently. You speak more directly. You move with less hesitation. And you don't have to maintain a pose. Your system is no longer locked around an old signal.

This process doesn't happen once. It's ongoing. Not linear, but layered. The more you work with the body as a source of memory — not intellectually, but somatically — the more access you have to emotional truth that doesn't depend on thought.

You stop managing symptoms. You stop controlling your state. And instead, you learn to participate — to feel what the body's already telling you. Clearly, quietly, and without resistance.

Movement, Release, and Self-Regulation

Movement is one of the body's oldest languages. Before words, before thought, the body moved to express, to defend, to connect, to process. It still does. Even in stillness, there's micro-motion: breath, pulse, shifts in balance. Movement is not optional — it's built in.

Yet modern life encourages stillness of the wrong kind. Not presence, but freezing. Not quiet, but suppression. People sit for hours, hold their breath unconsciously, disconnect from their own bodies. Emotion builds, but has nowhere to go. So it hardens into tension. Into posture. Into chronic activation of stress cycles.

Release is the body's way of returning to balance. But release doesn't always look how people expect. It isn't dramatic. It's not a breakdown. Sometimes it's as small as a tremble in the fingers, a spontaneous exhale, a subtle change in gaze. Other times it comes through full-body shaking, crying, or deep muscular unwinding.

This isn't performance. It's regulation. The nervous system, when allowed to move freely, knows how to discharge excess energy. But it needs permission — not from the outside, but from inside. And that permission starts with attention.

When you allow movement without control — not flailing, not forcing, but following the signal — the system starts to self-correct. You might stretch in a way that wasn't planned. Yawn unexpectedly. Shift weight. Arch your back. These movements aren't random. They're the body's way of reorganizing itself. Completing gestures that were interrupted. Finishing emotional arcs that were never fully expressed.

One of the most intuitive forms of this is spontaneous movement. In a quiet space, with eyes closed or unfocused, you simply let the body lead. No choreography. No goal. Just noticing: where does the impulse to move arise? What happens when I follow it — slowly, honestly, without commentary?

In the beginning, it can feel unfamiliar. The mind wants to jump in, to name, to control. But if you stay with sensation — not imagination — the movement begins to reveal pattern. You might find that your arms keep pulling inward, or your spine wants to sway, or your mouth wants to open. There's often emotion underneath. Not always dramatic, but real.

The point is not to express emotion. The point is to let the system unwind what it's ready to release. That's regulation. The system resets itself not through force, but through access — when enough space and awareness are present for it to do what it already knows how to do.

Regulation also happens through repetition. Small movements done with presence — rocking, pulsing, swaying — send signals of safety to the nervous system. When done rhythmically and with breath, these movements build coherence. They shift the system from sympathetic (fight/flight) to parasympathetic (rest/repair).

You don't need complicated practices. What matters is honesty. Movement that comes from the body's actual need — not imposed from the outside. This makes the work personal, embodied, and sustainable.

Self-regulation doesn't mean controlling emotion. It means allowing the body to feel and move without being overwhelmed. Over time, this builds capacity. You're able to stay present with more sensation, more energy, more intensity — without shutting down or acting out.

This has real consequences. You respond differently in conflict. You recover faster from stress. You notice tension before it becomes pain. You trust your internal cues. Not because you're mastering a technique, but because your system has learned how to return to center without panic.

There's also a social aspect. When you're regulated, others feel it. Your tone, timing, and presence change. You no longer project urgency. You no longer react from old pattern. This doesn't mean you're always calm — it means you're in honest contact with your own state, which allows others to meet you more fully.

Movement, release, and regulation are not side practices. They are core elements of conscious experience. Without them, all awareness stays in the head. You might think clearly, but the body stays braced. You might speak well, but your tone stays tight. You might look present, but the system is still running on defense.

Working with the body in this way doesn't remove difficulty. It gives you a way to meet it. With breath. With clarity. With motion. And with the kind of presence that doesn't need performance — just truth.

The Body as an Antenna: Sensitivity, Boundaries, and Coherence

The body is not just a tool for movement or a container for emotion. It's a sensor — a live, responsive field of perception that constantly reads internal and external signals. You could call it an antenna: tuned not just to survival, but to subtle shifts in energy, emotion, space, and rhythm.

Some people feel everything. Their system picks up tone before words are spoken, senses tension in a room, feels the difference between attention that's present and attention that's performative. This isn't imagination. It's sensitivity — often untrained, often overwhelming, but deeply real.

Others feel very little. Their system learned to numb, to override, to focus outward. This is not failure. It's a form of protection. When sensation is too intense or too confusing, the body shields itself. But even in that state, the antenna still works. It may be muted, but it's never off.

Sensitivity is not weakness. It's range. It's the ability to register information at different levels: physical, emotional, relational, spatial. But without boundaries, sensitivity becomes overload. The system receives more than it can process. Instead of clarity, there's confusion. Instead of connection, there's collapse.

This is why boundaries matter — not as defense, but as structure. Boundaries allow sensitivity to be useful. They give the system shape. They define what is yours to hold and what isn't. Without them, the body takes on other people's signals, other people's stories, other people's stress.

A boundary isn't a wall. It's a filter. It lets in what nourishes and keeps out what distorts. Some boundaries are physical — where you stand, how close you allow someone to be. Others are energetic — how much attention you give, how much you absorb, how long you stay in contact. Boundaries are not created with tension. They're created with clarity.

Coherence happens when the system is aligned. When your thoughts, emotions, posture, and presence match. You don't have to project anything. You don't have to defend yourself. You become internally ordered — not rigid, not controlled, but synchronized. Others can feel it. Your presence feels clean. There's less noise in your field. Less distortion.

When your body is coherent, your perception sharpens. You can sense what's yours and what isn't. You can feel when something's off — not as anxiety, but as signal. You don't need to analyze it. You just notice, and adjust. That's what antennas do. They receive, interpret, and orient.

This capacity can be trained. Through breath, through movement, through presence, through repetition. The more consistently you return to your own internal rhythm — not as escape, but as base — the more stable the antenna becomes. You stop living in reaction. You start receiving reality with less interference.

The goal is not to become hyper-aware or ultra-sensitive. The goal is to live in a body that's honest. A body that gives you clear feedback. A body that can open and close, engage and rest, feel fully and still return to center.

As the system becomes more coherent, sensitivity becomes an asset. You don't just feel more — you feel more accurately. You become less reactive, not because you've shut down, but because you're not thrown by every signal. You've learned how to stay centered without going numb.

This is not spiritual. It's biological. And it's essential if you want to live in contact with reality — not just your thoughts about it, but the actual movement of energy, relationship, and space.

Your body is not in the way. It's the way in. The way in to perception, truth, presence, and power that doesn't depend on image or performance. Just contact. Just coherence. Just enough honesty to stay in touch with the signal — and act from there.

Chapter 11: Quantum Models and Consciousness

There's a strange overlap between quantum physics and consciousness studies. Not because the math of quantum theory directly explains the mind — it doesn't. But because some of the questions they raise start to sound familiar: What happens when you observe something? Does the observer influence the outcome? What exists before a decision is made? What is the nature of reality when it's not being measured?

These questions, though born in laboratories and particle accelerators, resonate with the kind of things people ask when they start paying attention to awareness itself.

Quantum mechanics, as a field, is precise, mathematical, and extremely successful in predicting how particles behave at microscopic scales. But it also describes a world that behaves nothing like our everyday experience. Particles can be in more than one place at once. Outcomes remain undefined until observed. Systems appear to "know" about each other instantly, across space. And the act of measurement collapses potential into specific form.

For physicists, these results are part of a framework that works, even if it remains difficult to interpret intuitively. But for people exploring the nature of consciousness, the metaphors are hard to ignore. They sound a lot like the inner experience of being aware: multiple possible thoughts, collapsing into a single choice; a felt sense of interconnection; the way attention seems to shape perception; and the impossibility of separating the observer from what is being observed.

This chapter doesn't try to turn physics into psychology. That's been done badly too many times. Instead, it takes a clear look at what quantum theory actually says, how it's sometimes (mis)used in discussions of the mind, and where the parallels — when approached honestly — might offer something real.

The goal is not to prove anything. It's to hold open a space where science and direct experience don't contradict, but inform each other. Where the strange behaviors of matter at the quantum level invite a different kind of humility — not one that explains everything, but one that questions the assumption that we already know what consciousness is.

The chapter will unfold in four parts:

- **What Quantum Physics Actually Describes**
  A plain-language look at superposition, entanglement, uncertainty, and the role of the observer in quantum theory — stripped of mysticism and explained in terms of what science actually claims.
- **Consciousness as Observer: Metaphor or Mechanism?**
  Exploration of whether the observer effect in quantum theory has anything to do with subjective consciousness, and how that idea has been both misused and thoughtfully explored.
- **Quantum Theories of Mind: Between Speculation and Insight**
  A look at scientists and thinkers who have tried to build bridges — from Roger Penrose and Stuart Hameroff to Eugene Wigner and David Bohm — and the key questions they raise.
- **Working with Quantum Ideas Without Losing Ground**
  How to use the structure of quantum thinking — multiplicity,

potential, collapse, entanglement — as a useful lens for inner experience, while remaining grounded and clear.

This chapter is not the conclusion of the book. It's a reflection point. A way to step back and look at what it means to think, to perceive, and to act from the understanding that reality might be less fixed than it seems. Not to escape into abstraction — but to return to life with sharper attention.

What Quantum Physics Actually Describes

Quantum physics is the science that describes how matter and energy behave at very small scales — far smaller than atoms. At this level, the rules we're used to no longer apply. Objects don't have fixed positions, events don't follow simple cause-and-effect, and certainty becomes a mathematical impossibility.

At the heart of quantum theory are a few key principles. These are not spiritual ideas. They are part of a framework tested millions of times in experiments, often with incredible precision. Let's look at four of the most important ones:

Superposition In classical physics, something either is or isn't — a ball is on the table or it's not. But in quantum systems, a particle like an electron can exist in multiple potential states at once. It's not that it's in several places at the same time in a blurry way. It's that its state — location, momentum, even identity — is *undecided* until measured. This is called superposition.

What collapses that potential into a single state? According to the standard interpretation (called the Copenhagen interpretation), it's the act of measurement. Before being measured, the system exists in a range of possibilities. After measurement, it has a defined state. This doesn't mean that the observer *consciously causes* the collapse — just that the interaction between system and measuring device (or environment) forces one outcome to appear.

Uncertainty The uncertainty principle, discovered by Werner Heisenberg, states that certain pairs of properties (like position and momentum) can't both be known with absolute precision. The more precisely you know one, the less precisely you know the other. This isn't a flaw in instruments. It's a feature of reality at this scale. Quantum systems don't allow total certainty.

This idea goes beyond measurement — it suggests that reality itself is fundamentally indeterminate in some respects. Not everything is fixed and hidden waiting to be discovered. Some things are open, waiting to be decided through interaction.

Entanglement When two particles interact in a certain way, they can become entangled — meaning their states become linked. If you measure one, the other's state is instantly determined, no matter how far apart they are. Einstein famously called this "spooky action at a distance," because it seems to violate the limit of the speed of light.

But this doesn't allow faster-than-light communication, and it doesn't mean one particle sends a signal to the other. It just means their states are mathematically correlated. Once you measure one, the system as a whole has resolved — even if parts of it are far away.

Entanglement has been confirmed in many experiments. It's a real phenomenon, but what it means is still debated. Some physicists think it suggests non-locality — that reality is connected in ways that don't fit within space-time as we know it.

The Observer Problem This is where most confusion begins. In many quantum experiments, the result depends on how the system is observed or measured. That led some early physicists — including Niels Bohr and John von Neumann — to wonder if consciousness itself was involved in this "collapse" from possibility to outcome.

But modern physics does *not* claim that human consciousness directly causes quantum events. In most interpretations, the "observer" is any interaction with the environment — a measuring device, a photon, a particle. Consciousness isn't part of the equation.

That said, some physicists and philosophers have continued to explore whether there's a deeper link. Is the act of observation somehow *fundamentally linked* to awareness? That remains speculative — not disproven, but not proven either. It's one of the few places where physics brushes up against philosophy in a serious way.

This is what quantum physics *actually* describes: a world that is not fixed until measured, where uncertainty is built-in, where entangled particles behave as if they are connected beyond space, and where measurement plays a strange, central role.

These ideas are fascinating — but they are also precise. They describe systems at the microscopic scale. Applying them to thought, emotion, or consciousness requires care. Not because the ideas have no value there, but because the rules change when you leave the world of particles and enter the world of minds.

That's what the next section will begin to explore: where the metaphor of quantum physics can help us describe consciousness, and where the boundary between science and speculation needs to be respected.

Consciousness as Observer

The idea that "consciousness creates reality" has been repeated so often — especially in spiritual or self-help circles — that it's easy to forget how subtle the original question was.

In quantum physics, the observer plays a unique role. In many experiments, a particle doesn't settle into a defined state until it is measured. The system exists in multiple possible states (a superposition) until something — usually an interaction with another system — forces one outcome to appear.

This interaction is sometimes called the "collapse of the wavefunction." But what exactly causes that collapse? And is a *conscious observer* required for it to happen?

That question has sparked debate for decades.

The mainstream answer in physics today is **no**. Most physicists believe that collapse occurs through interaction with the environment, not through awareness. A Geiger counter, a photon, or even a wall can cause a system to collapse into a defined state. Consciousness isn't necessary.

Still, the fact that *measurement* causes collapse — and that the system behaves differently when unobserved — leaves room for deeper questions. What is measurement? What makes something an "observer"? And what happens, exactly, when the possibilities become actual?

Early pioneers of quantum theory were not shy about asking these questions.
– Niels Bohr, one of the founding figures, emphasized that quantum systems don't have definite properties until they are observed – though he didn't claim that human consciousness caused this.– John von Neumann argued that the only thing that can't be modeled physically is consciousness itself, so it must be the final link in the chain of measurement.– Eugene Wigner, another respected physicist, suggested that consciousness might play a fundamental role in causing collapse – though even he acknowledged this was speculative.

The idea that consciousness has a central role is called the "observer effect" interpretation, and it's still debated. Not because it's proven, but because *no one fully agrees on what measurement really means*. That's what keeps the question alive.

Some interpretations (like many-worlds) say there is no collapse at all — that every possible outcome happens, in parallel, and we just experience one. In that model, the observer doesn't cause anything — they just experience one path out of many. Consciousness is just along for the ride.

Other approaches, like objective collapse theories, suggest that collapse happens due to physical thresholds — not consciousness. And in Bohmian mechanics, particles always have real positions; the uncertainty comes from not knowing the "hidden variables" that guide them.

So where does that leave consciousness?

At this point in science: outside the theory. Quantum physics, as it's used today in labs and equations, doesn't include consciousness as a variable. It describes systems with remarkable precision without needing to invoke the mind. But that doesn't settle the deeper philosophical question — it just avoids it for now.

Still, for many who work with inner experience, the quantum model remains *compelling* — not because it proves anything about the mind, but because it mirrors certain subjective experiences:

- That possibilities feel real before decisions collapse into action.
- That attention seems to affect what we notice, and what grows in focus.
- That intention feels like it shapes outcome, even subtly.
- That observation itself carries power — not control, but influence.

These parallels don't prove a mechanism. But they suggest a metaphor worth considering. Perhaps consciousness doesn't *cause* collapse — but maybe it participates in reality in a way science hasn't yet modeled.

The risk is always the same: confusing metaphor for mechanism. Saying "consciousness is like an observer" is useful. Saying "consciousness creates reality" as if it's settled science is misleading. The truth lies somewhere in between — where language starts to break down, and both physics and philosophy begin to listen more than declare.

In the next section, we'll look at thinkers who have tried to build real models — bridging quantum theory and mind. Some stayed within the bounds of science. Others stepped into speculation. What matters is how they approached the question — not with dogma, but with curiosity, and with the humility to admit we don't yet know how deep the connection goes.

Quantum Theories of Mind

Most scientists avoid linking quantum physics with consciousness. The two fields operate on vastly different levels: one measures particles, the other explores awareness. But a handful of thinkers have tried to build bridges. Some with care. Some with boldness. And some with ideas still debated decades later.

The most well-known attempt to create a direct quantum model of consciousness is the Orch-OR theory, proposed by Roger Penrose and Stuart Hameroff. Their model is ambitious. It suggests that quantum processes happen inside the brain — not just anywhere, but specifically in the microtubules inside neurons, structures previously thought to be too warm and chaotic to support quantum behavior.

Penrose, a physicist and mathematician, introduced the idea of objective reduction — that quantum superpositions collapse on their own, based on the geometry of space-time. He suggested that this process might be linked to consciousness, if it occurred in the right biological environment.

Hameroff, an anesthesiologist, added the biological framework. He argued that microtubules could support quantum coherence — meaning they could maintain quantum states long enough to affect neural processing. If true, that would mean parts of the brain might operate not just through chemistry and electricity, but through quantum dynamics.

The Orch-OR model is elegant, but controversial. Critics argue that the brain is too "noisy" and warm for quantum coherence to last long enough to influence anything. Experiments in quantum computing rely on near-zero temperatures and total isolation. How could a living brain — messy, active, biochemical — maintain that kind of precision?

Supporters respond that recent research in quantum biology has shown surprising effects — like quantum coherence in photosynthesis and bird navigation — happening in warm, wet conditions. So the idea isn't impossible. Just unproven.

Other thinkers have proposed different models. Eugene Wigner, a Nobel laureate in physics, believed that consciousness might be a fundamental part of the measurement process. He argued that collapse doesn't fully happen until a conscious observer registers the result. That idea has mostly fallen out of favor in mainstream physics, but it still lives in philosophical circles.

Henry Stapp, another physicist, continued this line of thought, proposing that the mind interacts with quantum fields in ways that guide neural behavior. His work is speculative but deeply informed by both physics and philosophy.

Then there's David Bohm, who proposed a very different kind of quantum theory — one that includes an implicate order: a deeper level of reality where everything is connected, and what we see is just the unfolded surface. Bohm believed that thought itself might be part of this larger process. He spoke of mind and matter not as separate, but as different expressions of the same underlying field.

Bohm didn't offer a mechanistic theory of consciousness. But he did offer a way of thinking: that fragmentation — the way we divide self and world, mind and matter — might be an illusion, caused by limited perception. His work inspired dialogue between physicists, philosophers, and even spiritual teachers.

There are also modern neuroscientists and theorists exploring quantum-inspired models — not because they believe the brain is literally a quantum computer, but because classical models of brain function don't fully explain subjective experience, or the unity of consciousness. Some explore quantum information theory as a new lens. Others look at entanglement as a metaphor for the kind of binding that happens in attention, memory, or relational presence.

Still, most of these models remain on the fringe — not because they lack imagination, but because the tools to test them are not yet available. The risk in all of them is the same: crossing the line from bold theory into unfounded claim. But the reason they persist is also clear: no current model of the brain fully explains consciousness.

We don't know how awareness arises. We can track neural activity. We can study behavior. But we can't explain why an electrical signal in the brain *feels* like anything. That mystery — known as the hard problem of consciousness — remains unsolved.

Quantum models don't solve it either. But they remind us that the universe may be stranger than we think. That there might be levels of order — or interaction — we haven't yet learned to detect. And that between particle and perception, there's still a space science hasn't mapped.

Working with Quantum Ideas Without Losing Ground

Quantum physics has a particular kind of gravity. Its concepts are counterintuitive, precise, and powerful. But when taken out of context, they're also easily misused. Terms like "frequency," "field," "vibration," and "observer" have spread into popular language, often detached from what they actually mean.

Still, the core ideas behind quantum theory remain useful — not as scientific claims about the mind, but as metaphors that can support insight, reflection, and awareness. The key is to use them with clarity, not to decorate language, but to deepen attention.

Superposition can be seen as a model for the internal state before decision. When you hold multiple possible choices, emotions, or interpretations — all present but unresolved — that's not confusion. It's potential. You haven't collapsed into one direction yet. When a decision is made, when clarity arrives, the field narrows. You move.

In this way, you can work with your own uncertainty not as a flaw, but as part of a natural process. Instead of rushing to collapse every question, you can stay with the open state a little longer — allowing more information to appear, allowing the deeper signal to emerge. Sometimes what's needed is not more thinking, but less interference.

Collapse of the wavefunction is often misused to suggest that thoughts instantly shape physical reality. That's not how the physics works. But subjectively, it's not meaningless. When you give your full attention to something — a decision, a relationship, a gesture — you're creating structure. You're moving from possibility into form.

So attention matters. Not in a magical sense, but in a mechanical one. What you focus on tends to grow. What you act on becomes real. Not because you're altering particles with your mind, but because you're participating in life with presence instead of drift.

Entanglement is often used metaphorically to describe emotional or energetic bonds — and while it doesn't imply literal quantum entanglement between people, it does point to something real: the way internal states can become linked. In deep connection, attention becomes shared. Change in one person seems to affect the other. Not as telepathy — but as resonance, timing, and pattern.

This is not fantasy. It's pattern recognition. When two systems interact long enough, they begin to coordinate — in biology, in conversation, in emotion. The nervous systems of two people can sync heart rate, breathing, even micro-movements. This isn't quantum behavior, but it is non-linear — and deeply human.

Uncertainty, in quantum mechanics, is not just a gap in knowledge. It's a built-in feature. In experience, it can be a reminder: not all things need to be known to be real. You don't have to name every feeling to feel it. You don't have to explain every reaction to listen to it. You don't have to resolve every contradiction to act with clarity.

Bringing these ideas into inner work means dropping the need to explain everything. It means allowing space for what's unclear — not as avoidance, but as respect. Some states don't need immediate interpretation. Some signals take time to organize. The nervous system knows this. It has its own timing, its own rhythms of insight and integration.

What matters most is the posture you bring to these ideas. You're not trying to become a quantum being. You're not trying to turn physics into magic. You're simply using a set of models that reflect something you've already felt: That attention creates change. That presence shapes outcome. That clarity isn't always linear. That reality responds differently when you meet it with real contact.

So use these concepts carefully. Not as slogans. Not as explanations for everything. But as tools that remind you how to move — with awareness, with precision, and with the willingness to stay honest in the unknown.

There's no need to pretend these ideas make you more powerful. What they can do – if held correctly – is make you more precise. More humble. More able to sit in the spaces where thought runs out, and attention begins.

That's where science and consciousness briefly touch. Not in answers, but in the kind of questions that change how you see.

Conclusion

We live inside structures we rarely question. Thought, identity, memory, belief – all feel solid, until something breaks the pattern. Then, for a moment, something else becomes visible. A kind of spaciousness. A sense of presence that isn't based on concept or language.

This book wasn't written to give answers. It wasn't written to prove that the mind controls reality, or that consciousness is a quantum force. It was written to point toward something many people feel but don't always name: that perception isn't fixed, that inner states matter, and that the way we pay attention shapes what becomes real – internally, relationally, and sometimes far beyond.

The quantum references weren't meant as decoration. They were included because they offer an unusual invitation: to recognize that even in science, certainty has limits. That not everything is observable in a straight line. That interaction changes outcome. And that in some layers of reality, the observer is not separate from the observed.

But most of this book has stayed close to direct experience: attention, emotion, body, thought. That's where clarity begins. Not in abstract belief, but in the way you stand. In the way you breathe when no one's watching. In the way you speak when it matters. In the way you return to yourself after confusion or noise.

Power, here, is not about control. It's about contact. With sensation. With signal. With what is actually happening. That's the shift. Not becoming someone new. But dropping what's false, and moving with more precision, more honesty, more coherence.

You don't need to live in altered states to change. You don't need to be mystical to be real. You just need enough awareness to stop repeating what isn't true, and enough presence to choose what is.

There is no final method. No perfect language. No system that replaces the need to pay attention. There is only this: the moment you become aware of your own state, you're no longer trapped in it. From there, something else is possible.

That's not philosophy. That's perception.

And it's available now.

About the Author

The author has spent over two decades studying the intersection of attention, perception, and the non-linear dynamics of inner experience. With a background that spans neuroscience, somatic practice, and systems thinking, their work moves beyond theory — grounded in direct observation and applied awareness.

Rather than aligning with a single tradition or methodology, their approach is integrative: shaped by real-time interaction with the body, mind, and field. Influenced as much by silence as by science, they work with individuals and groups to refine attention, release habitual tension, and restore coherence in both perception and behavior.

They do not teach beliefs. They offer tools. The emphasis is on clarity — not as an idea, but as a skill. Whether working with altered states, emotional regulation, or decision-making under pressure, the focus remains consistent: clean perception, honest action, and internal alignment without illusion.

This book is the result of years spent in direct contact with process — not from the outside, but from within. It is written for readers who are ready to go beyond language, beyond self-image, and into the real structure of experience.

www.ingramcontent.com/pod-product-compliance
Lightning Source LLC
LaVergne TN
LVHW022100200926
848481LV00003B/1006